THE MILLFIELD MINE DISASTER

THE MILLFIELD MINE DISASTER

RON LUCE

Published by The History Press
Charleston, SC
www.historypress.com

First published 2024

Manufactured in the United States

ISBN 9781467155410

Library of Congress Control Number: 2023945944

To those who died on November 5, 1930, and those who loved them.

CONTENTS

LIST OF ILLUSTRATIONS

Graph and Tables

Photographs

Illustrations

ACKNOWLEDGEMENTS

Prior to 2010, I never imagined that I would find myself writing anything about coal mining or the labor movement in southeastern Ohio. I had come to southeastern Ohio in 1980 to attend Ohio University, where I focused on teaching writing composition and on American literature. I fully expected I would go elsewhere after graduating. However, like so many people who come to the area, I fell in love with it and ended up staying, finding work at a nearby college and getting involved with the community. I engaged in community theater and started doing some acting for training programs, where I had the good fortune to work with Kathy Devecka, a voice-over artist and voice/speech and educational theater consultant. In addition to theater for training purposes, Kathy was involved in historical reenactments and encouraged me to get involved. Soon, she had me learning everything I could about Thomas Worthington (one of the early leaders in promoting Ohio statehood and the state's sixth governor) and doing living history performances. Through her, I was introduced to John Cunningham, who is an exceptional genealogist and a walking encyclopedia of local history. It didn't take long before I became involved in gaining a much wider understanding of early pioneers, the development of communities, the history of enterprises and major historical movements that are associated with the area—particularly coal mining.

Kathy, always the enthusiastic promoter of living history, introduced me to John Winnenburg and Cheryl Blosser, who were the leaders of the Little Cities of Black Diamonds project and who were promoting tourism and historical awareness of the role of southeastern Ohio coal miners in the early days of the labor movement. I had no idea that the foundations for the United Mine Workers of America were at least partially built on the work that had been done by people in southeastern Ohio, particularly those living in New Straitsville and Shawnee. John and Cheryl, aided by Kathy Devecka, encouraged me to take on the character of Christopher Davis, a union promoter who worked diligently for all miners—but especially those in southeastern Ohio—to obtain fair pay and safe working conditions. When I said yes, I knew I would have to learn about not only the man but also his profession as a coal miner and labor leader in enough detail to be able to answer questions from knowledgeable audience members who like to see if the living history actor really knows what he's talking about. I played the role successfully many times.

When I retired from academia, I knew that the Athens County Museum and Historical Society was looking for a director. I applied and somehow managed to convince the board members who were doing the hiring that they should give me the opportunity. As a result, I learned about many aspects of local history that I had not known previously. I no longer recall what triggered my interest in the Millfield mine disaster in particular, but whatever it was, it has stayed with me for many years now.

This book has taken a long time to come to fruition. I am very grateful to Ms. Melody Bragg, who was, in 2011, a technical information specialist at the Technical Information Center and Library, National Mine Health and Safety Academy. She took significant time to copy and send the entire final report of the committee that investigated the 1930 disaster. She sent pictures, telegrams, letters, an early draft of the final report—a treasure trove of information. The materials provided the necessary foundation for the research done since that time. Over the years since she sent the materials, I have used what she sent in many presentations. I was particularly happy to have the data and findings of the investigative team—people who (1) were actually at the mine shortly after the disaster, (2) interviewed countless people in order to compare and contrast versions of what happened, (3) did underground research work to study physical evidence and (4) put their findings in coherent written forms.

When John Rodrigue of The History Press sent an email asking recipients if they had ideas for new books, I responded almost immediately to tell him I would like to write about the disaster. He presented my proposal to the company's decision-makers and gained their approval for me to begin the process of writing and publication. John played the role of mentor and provided me with the resources I needed to complete the publication process, all with kindness and professional regard.

The work of the investigators Ms. Bragg forwarded to me was particularly important for my research; from my earliest experiences trying to understand what happened at the mine, I learned that looking into newspaper accountings to tell this story is, to say the least, maddening due to conflicting information (sometimes within a single article), as well as confusion about details and names of people involved. I have tried to use newspaper information sparingly. I would agree with Daniel Harrington, chief engineer of the Safety Division, United States Bureau of Mines, who wrote in a letter to J.J. Forbes on November 28, 1930:

> *The clippings* [newspaper articles] *have in them a considerable amount of interesting information, and also a considerable amount of pure "bunk."*

Jim Mingus, who is the grandson of Oscar and Minnie Willis, has been gracious in allowing me to use the pictures of his family members, three of whom died in the mine on November 5, 1930: his grandfather Oscar and two uncles (sons of Oscar), Andrew and Virgil. In addition to photographs, Jim shared stories about the devastation caused by the losses and the effects on the remainder of the family. Through Jim, I had the good fortune of meeting his mother, Kathleen (the youngest in the Willis family), before she died and talking with her about the disaster, her recollections and the effect it had on the family.

Cyrus Moore III, a historian and friend who has expertise in Ohio military history, was most gracious in providing me with clarification about newspaper reporting that referred to "militia," "the Athens Militia" and the "Ohio National Guard." The terms seemed to be used interchangeably, yet I knew there was at one time an Athens Militia and, of course, an Ohio National Guard. Out of a desire to be accurate about who was actually called in to assist at the mine, I contacted him and asked him to help me understand. He explained:

> *By 1930, the Athens Militia was part of the Ohio National Guard (ONG). All enrolled militia companies were part of the state organization under the command of the adjutant general. The only independent militia still in existence was the Cleveland Greys. The general public at that time was still accustomed to calling the ONG "the militia"...the Athens Militia at the armory was a battery of the 134th Field Artillery, ONG, at the time.*

Valuable research assistance was provided by Thomas McAnear from the National Archives at College Park, Maryland; Todd Crumley, still picture reference, National Archives and Records Administration; George Franchois, director of the U.S. Department of the Interior Library; Andrew Tremayne, Heritage Program, Forest Service, Wayne National Forest; Beth Wilson, records coordinator, and Brent Heavilin, permitting and bonding manager, Ohio Department of Natural Resources; Scott Keyes, chief of Heritage Documentation Programs, National Park Service; and Laura Smith and Janet Carleton at the Mahn Center for Archives and Special Collections, Ohio University.

All writing is enhanced by the contributions of an excellent editor and talented staff members who bring a book to life. Ashley Hill, my editor, and all of the workers at The History Press have my deepest appreciation for the amazing work they do.

And now I return to John Cunningham, who has spent many hours assisting me with genealogical research related to the eighty-two men who died on November 5, 1930, and with verifying facts about locations and people in and around Millfield. Together, we checked and double-checked the names of the dead, trying to bring pieces of their lives together and give them the honor and dignity they deserve by recognizing a few facts about each of them: birthdate, parentage, marital status, children in the home, military service, burial date and burial location, as well as the death certificate number—our goal being to make it easier for anyone wishing to do research about the victims in the future. Regrettably, we were unable to do equal service for some of the men who came from other parts of the world. They often took on names that disguised their foreign-born identities and left very little for us to work with in our efforts to honor the lives they lived. I can only hope that someday, someone will find a way to unlock their secrets.

Thank you, one and all!

1

OHIO'S WORST MINE DISASTER

It was Wednesday, November 5, 1930, when 82 men lost their lives in Millfield, Ohio, as a result of an explosion that ripped through the underground workings of the Sunday Creek Coal Company Mine no. 6. In an instant, 56 women became widows and 137 dependents found themselves fatherless. Millfield became a focal point for local, state and national news, and another chapter in the history of coal mining in Ohio and the United States was written.

Before approximately 11:45 a.m. that day, Millfield had been just another small town in Athens County where people struggled to survive, largely by working the land as farmers or by working in the nearby coal mines. Very few people in the immediate area came close to what might have been thought of as the middle class at that time. The United States was reeling from the onset of the Great Depression launched by the stock market crash of 1929. Jobs were scarce, supplies were short and money was not readily available. However, on that day, when the sounds of an earthshaking blast shot up out of Mine no. 6 and fell like shrapnel against the ears of people aboveground, all thoughts and fears about the situation of the United States and world economies and anticipated woes based on the Depression vanished.

People aboveground at the mine site and at the new fan station approximately a mile and a half away from the main shaft knew that whatever had happened was likely to have been devastating to those working underground. A blast of that magnitude almost certainly had injury and death accompanying it. At the new airshaft, a local person happened to be

Author collection.

present as a man who had been working on the fan was blown off the housing and thrown at least fifteen feet away from it; in fact, that particular event triggered the first calls to mine officials and local authorities. Then calls were made to state and national mine officials to ask for assistance and rescue equipment. The perceptions that there would likely be deaths involved were strengthened when miners who had survived the blast started coming up through the main shaft, some of them injured or feeling the effects of the poisonous gases they had breathed before escaping.

The blast triggered the worst fears of family members and friends who had run or otherwise found their way to the mine site hoping against hope that no one, especially their loved ones, had been killed or seriously injured. As news spread, many others began to arrive: volunteers from the Red Cross; members of the Ohio National Guard; state and national mine bureau representatives; local and regional doctors, nurses and undertakers; numerous people who wanted to help somehow; news reporters; and hordes of those who might be best described as "merely curious." All of them gathered throughout the day at the mine's aboveground properties. Early on, they came in the hope there would be survivors; within a few hours, however, most knew that, short of a miracle, they would be spending their time waiting for the dead to be brought out of the mine.

The Millfield mine disaster is, to this day, the worst mining disaster in Ohio's history. The event is embedded in the culture of southeastern Ohio and in the minds of family members who for generations afterward have wondered why it happened, whether it could have been prevented and who might have been to blame. Though the Millfield Mine no. 6 explosion is far from the deadliest mining accident when compared to those that occurred in other states or other parts of the world, it provides a historical reminder of the inherent dangers of coal mining and is pivotal to Ohio's history as a major player in the labor movement in the United States. Behind the story of Sunday Creek Coal Company's Mine no. 6 on November 5, 1930, is another story: the story of long-standing struggles between workers and owners regarding fair pay and

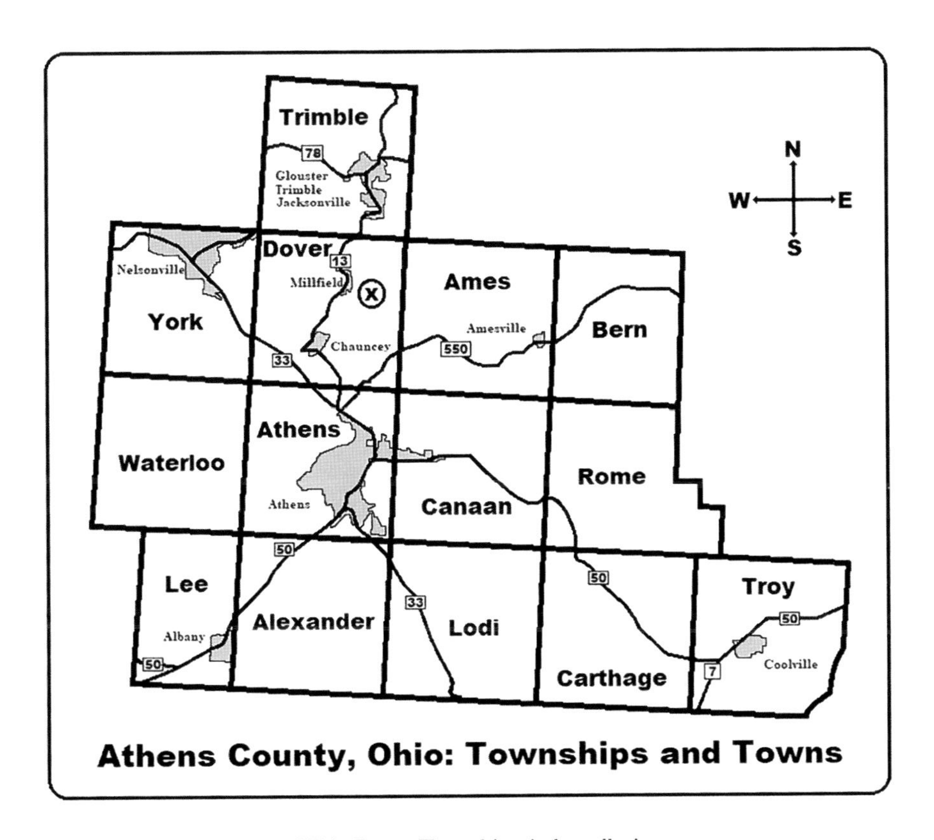

Location of mine site: see "X" in Dover Township. *Author collection.*

safe working conditions. It is a story of corporate risk-taking and cutting corners at the workers' expense to maximize shareholder profit.

Too often, in the writing of historical events, the focus is on the what, when, where, why and how of the event; the "who" is often reduced to talking about someone in a leadership role or someone who performed a heroic or villainous act or someone who had an exciting life story that will resonate with readers. Others who were involved are often bundled together and labeled with words such as *the dead*, *miners* or *workers*. It is important to remember that eighty-two men ranging in age from seventeen to sixty-nine—men who loved, laughed, cried, dreamed, hoped and did their best to survive and care for their families in the early years of the Great Depression—had their own lives worth living and worth recording accurately as part of history.

Throughout this book, every effort has been made to exhibit as much of the documentation as possible that was provided by people who were on-site in the aftermath of the explosion. Reading their statements based on professional experience and having been on-site provides the best information available. Though an author influences readers by the choices of material selected for use, the organization of those materials and language choices, the goal of this author has been to allow the reader to absorb disparate pieces of information that provide facts to make sense of disparate information and then test their own responses against the author's conclusions.

Many years have passed since the explosion that took the lives of eighty-two men and changed forever the lives of those they left behind. Anyone who might have had some information to contradict or confirm the records available for this book is long gone. What remains beyond the stories, myths and misinformation still being passed down from one generation to another is what was put into print in letters, telegrams, reports, newspapers and other researchers' renditions of the event.* To date, no one book has attempted to pull together all of the known factors involved in the Millfield mine disaster, including a history of the mine itself, the relationship of the Millfield incident to the larger issues at the heart of the labor movement, the actions taken at the time of the disaster and a definitive list of the men who died. In the text that follows, this author attempts to rectify that situation.

A Brief History of Coal Mining in the United States

Fully understanding what happened at Millfield in 1930 requires some background in the history of coal mining. First of all, coal mining has been around for a very long time. Andrew Roy (1834–1914), Ohio's first chief mine inspector, wrote in his book *A History of the Coal Miners of the United States* (n.d.), "Coal was mined along its outcrop in the British isles with picks made of oak and flint before the birth of Christ."[1] The Romans, upon conquering Britain, were known to have taken prisoners of war and condemned them to live and die in underground mines.[2] For centuries, coal mining in the British Isles provided opportunities for miners to learn the craft of mining for maximum efficiency, the ways that owners and operators used power to control and manipulate miners to risk their lives to make the owners wealthy and how unionization appeared to be the only means by which change might occur.

* Douglas Crowell has done admirable work telling the basic story of the disaster, and his publications are excellent. However, no one has published the wide-ranging factors and explanations established in this book.

Many early immigrants to the United States, especially those coming from England, Scotland and Wales, brought vast knowledge of coal mining methods from years of experience as miners. They were highly sought after in the United States for leading and overseeing workers in mines. Andrew Roy was born in Scotland. Early miners and leaders included people like Daniel Weaver (England), Thomas Lloyd (England), John Siney (Ireland/England) and Christopher Evans (England/Wales). These men arrived in the United States with a commitment to the concept of unionism to provide a fair system for the distribution of wealth so miners could properly care for themselves and their families; they encouraged American miners to join forces for achieving fair wages and safe working conditions.

> *These men were the best miners in the world. Thousands of them... emigrated to the United States. They had been the recipients of the benefits of labor combinations in the mother country in the way of shorter hours, increased safety, better ventilation, educational advantages, and larger wages, and had become as strongly attached to the necessity and value of workingmens'* [sic] *unions for mutual protection as they were confirmed in the belief of a future state. They spread the gospel of union among American miners with great zeal and enthusiasm.*[3]

The existence of coal in what is now the United States was known as early as 1669. The first American mining company was the Lehigh Coal Mine Company (Pennsylvania), formed in 1803. As cities grew and wood for heating people's homes grew less plentiful, coal provided an alternative heat source. It was also an essential ingredient in the complex process of creating coke, which then could be used for creating the intense heat required for making iron and steel. The first beehive coke ovens in the United States were built in Connellsville, Pennsylvania, in 1841. Coke is known to have been made in Ohio in places like Utley and Leetonia, and it was likely made in many other locations within the state as well.

With the arrival of the railroads came new coal-hungry machines—trains—which, in turn, created a vast new network of clients and profits for coal companies. Coal companies worked hand in hand with the railroad companies to bring spurs (special sets of tracks) right up to company tipples to collect coal and cart it off to new markets or to current markets more rapidly, thus reducing or eliminating wagon hauling to waterways for long-distance shipping. Train transport brought coal to some markets in less than a day—as opposed to weeks of shipping via waterways (canals, rivers, et cetera). Access to railroads aided in the development of new mines as well. Wealthy investors and savvy entrepreneurs with access to rich clients quickly bought up huge quantities of acreage and mineral rights throughout Pennsylvania, Ohio, Kentucky, Virginia (in the portion of the state that later became West Virginia) and, later, throughout the United States as the country expanded westward and new, rich deposits were identified. In effect, coal was a major factor for myriad changes throughout the United States.

> *In ten years from the time the first shipments were made by rail, the annual* [coal] *output had risen to nearly 6,000,000 tons, giving employment to 10,000 miners and mine laborers. In 1853, the output had reached 11,000,000 tons, and in 1873...there were 22,880,000 tons mined, giving employment in round numbers to 50,000 workers.*[4]

Railroad and coal companies learned quickly that their destinies were bound to each other. Coal delivery was a major source of income for most of Ohio's railroads, and railroads were the most cost-effective means of delivering the products of the mines. By working together, owners and operators strived to maintain competitive pricing with other railroad lines and mining companies vying for access to greater wealth. Like many other businesses, the railroads and the coal companies were susceptible to fluctuating market and supply demands, which sometimes led to the need to raise or cut costs in order to stay in business or to maintain the level of profit the companies' stockholders had become accustomed to. A major strategy for maintaining high profits throughout the world of commerce—but particularly in coal mining—was to use employee wages to adjust for any losses or reductions in customer orders. Companies had used this approach since the early days of coal mining in the United States.

It was not uncommon for miners who were being paid one rate per ton of coal they produced to be told that the rate would be cut by a percentage, sometimes effective immediately after announcing the cut. The owners and operators seemed to have worked from the belief that if miners objected, they were free to leave; there would always be others who could take their place for the wages the owners were offering. This was true. One of the biggest problems miners faced was that there were many people who would take their jobs for less money than the miners were making. Often, there were far more people wanting to work than there were jobs for them. Needing to feed their families and put roofs over their heads, people out of work convinced themselves that it was better to have any pay than no pay. Of course, once they got the jobs, they learned quickly that the low pay was insufficient to meet their needs, and they would be forced into debt through the necessity of accessing credit at the company store. Essentially, employees and people who wanted to become employees were working against one another and their own interests. Many owners, operators and mine managers took full advantage of such situations to keep pay low and profits high. Adding to the complexity of unionization was that few miners had the luxury of just walking away. Going elsewhere was not always an option, and finding another job in mining couldn't guarantee that pay wouldn't be cut or that a company wouldn't succumb to reduced demand for coal.

Other factors playing into the difficulties of mine work were the social implications of seeking better pay. There were always the issues of whether there would be work elsewhere and then dealing with the difficulties and costs involved in moving—transportation, carting household accumulations, finding accommodations, giving up the rituals built within a community over time, pulling children away from schools and friends and ending pleasurable relationships with other families. It was often easier to live up to the notion of "better the devil you know than the devil you don't."

It was not uncommon for miners to find themselves deeply in debt to the companies they worked for. In many cases, coal companies in the United States provided rental housing for miners and their families and took the rent out of their pay. Company stores provided miners and their families access to food, clothing and household needs on credit. Unfortunately for the miners, company-owned stores often sold goods at prices that were higher than what would have been available in nearby towns. However, shopping elsewhere required actual cash. For many years, miners were paid in scrip (tokens) that could be used only at the company store run by the company they worked for, and when the token scrip ran out—largely because miners weren't paid sufficiently to cover the costs of caring for their families—the miners had little choice other than to buy on credit. (This is where the phrase "I owe my soul to the company store" comes from.) If miners complained about their pay, work conditions or prices at the company store, they risked raising the hackles of the foremen or owners, who would then put them out of their homes, fire and blackball them or punish them with work assignments that made it difficult for them to make even minimum pay, thus putting them further in debt.

Sometimes, work stopped altogether, leaving miners with few means of survival other than buying on credit or finding temporary work—if any could be found. Coal production had always depended on orders from external clients. These clients made buying decisions based on their current and projected needs, weather conditions, prices and, in some cases, the marketability of their own products requiring coal as part of their manufacturing processes. A mild winter could dramatically affect the demand for coal in some areas. A sudden shortage of coal cars for transportation by rail could also cause a company to stop production until the railroads could resolve the problem. A competing coal company could entice customers away with cheaper prices, and production might have to wait until new customers could be obtained. Slowdowns in production were not uncommon. A few weeks or months of slowdowns in coal production meant miners might work very little or not at all during those times, making it difficult or impossible for them to pay their debts or support their families without asking for more credit at the company store. When work in the mines started up after a slowdown, miners' debts would be deducted from their pay, leaving very little for the daily expenses of living and supporting their families' needs. Even after the days of scrip, it was difficult for miners to manage their low pay around boom or bust work opportunities or changes in pay scales dictated by owners.

Though dealing with the finances of their lives as miners was difficult, it didn't compare with the dangers of the work itself. Thousands of miners died in the mines or as a result of their various experiences working in the mines: accidents, explosions and exposure to poisonous gases and diseases, such as coal workers' pneumoconiosis (CWP), commonly known as "black lung disease." The Department of the Interior, Bureau of Mines, Bulletin 115 provides statistics that show that between 1870 and 1900, 17,640 people died in mines, of those, 1,860 died in explosions.* The United States Department of Labor Mine Safety and Health Administration's statistics show that there were 104,915 mine fatalities between 1900 and 2022, with the majority occurring before 1950. Although there are no official counts of mining deaths prior to 1870, it does not require much reading about mining to guess that deaths in the U.S. mines before that date were also numerous. Extracting data from Hiram Humphrey's "Historical Summary" shows that approximately 2,600 miners died from explosions between 1881 and 1950. The following tables provide quick snapshot views of how dangerous mining has been nationally as well as in the state of Ohio:

* This data is "for inspection states only" and should be considered a minimal assessment.

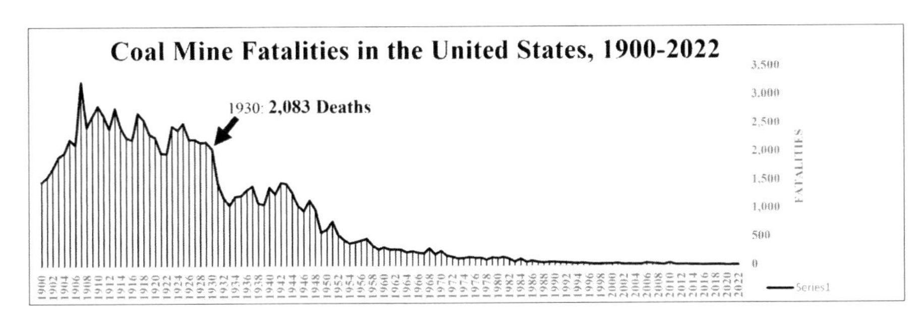

Coal mine fatalities in the United States, 1900–2022. *Data from the United States Department of Labor, Mine Safety and Health Administration.*

Table 1: Deaths and Serious Injuries in Ohio Mines by Calendar Years 1874–1915

Year	Deaths	Injuries	Year	Deaths	Injuries
1874	20	80	1892	42	93
1875	23	40	1893	32	123
1876	13	70	1894	45	116
1877	30	39	1895	52	152
1878	20	x	1896	41	159
1879	x	x	1897	40	142
1880	22	61	1898	52	155
1881	29	x	1899	59	204
1882	25	25	1900	68	207
1883	26	x	1901	114	324
1884	26	40	1902	81	298
1885	51	32	1903	124	308
1886	43	65	1904	118	316
1887	36	75	1905	114	372
1888	29	79	1906	127	384
1889	33	52	1907	153	493
1890	42	52	1908	112	426
1891	44	66	1909	115	467

Year	Deaths	Injuries	Year	Deaths	Injuries
1910	161	471	1913	162	532
1911	109	375	1914	61	x
1912	136	395			
Sub Totals:	**631**	**1,108**		**1,999**	**6,180**

Fatalities = 2,630 **Serious Accidents = 7,288**
Avg. deaths per year = 66 **Avg. serious accidents per year = 202**

x = no data available

Deaths and serious injuries in Ohio mines by calendar years 1874–1915. *Data from available Ohio Annual Reports of the Chief Inspector of Mines. (After 1913, mine inspections came under the auspices of the Ohio Industrial Commission).*

Table 2: Explosions in Ohio Mines 1881–1950

Explosion Site	Mine	Date M-D-Y	Killed
Robbins	Robbins	02-10-1881	6
Amsterdam	Amsterdam	04-21-1910	15
Belle Valley	Noble	05-17-1913	15
Millfield	Sunday Creek # 6	11-05-1930	82
St. Clairsville	Willow Grove #10	03-16-1940	72
Byesville	McFarlin	12-28-1942	3
Belle Valley	Belle Valley #1	01-10-1943	3
Bergholz	Wolf Run	05-29-1943	1
Barton	Barton	09-05-1944	0
Roseville	Launder	11-18-1947	1
		TOTAL	**198**

Explosions in Ohio mines 1881–1950. *Data extracted from Hiram Brown Humphrey's book* Historical Summary of Coal-Mine Explosions in the United States *(Washington, D.C.: United States Department of the Interior, Bureau of Mines, 1959).*

Miners risked their lives every time they went underground. Slate falls, mine collapses, poisonous gas inhalations and explosions were always potential risks.

Once coal-cutting and hauling machinery was introduced into mining, new dangers, such as machine failures and machine-caused injuries, came into being. Given all of the death and destruction, it is no wonder that miners continually asked their employers to provide as much safety protection as possible. It is even less a wonder that they felt they should be fairly paid for the difficult work and dangerous working conditions they faced. Unfortunately, their needs were not always high on the priority lists of mine owners and operators.

Unionization has long been the only means by which miners could make changes in a system that favored wealthy owners over workers. The collective power of unionization gave miners their only means of countering the seemingly limitless resources that were used against them. All attempts to bring unionization into existence were met with fierce opposition from owners and operators who formed syndicates (not all that different from unions) to undermine and destroy the workers' efforts to gain any power over them. Given the power and wealth used to destroy unionization, it is not surprising that a number of unions came and went over the years: the Knights of Labor, Ohio Miners Amalgamated Association, the National Federation of Miners and Mine Laborers, the National Progressive Union of Miners and Mine Laborers and others.

The need for unionization under a powerful organization was not abandoned by miners, despite their many defeats. A powerful demonstration of the miners' determination can be found in the example of Ohio's 1884–85 mine strike—a major confrontation between miners and mine owners. At that time, Ohio miners were unionized as the Ohio Miners Amalgamated Association (OMAA). Union representatives were attempting to reason with owners and operators who wanted to cut worker pay. Prior to January 1884, miners were being paid eighty cents per ton of coal. The syndicate (a coalition of owners/operators) and the Ohio Coal Exchange put forward a call for a twenty-cent cut in pay to go into effect on March 1, 1884. Workers objected and asked the companies reconsider. After negotiations, the syndicate and Exchange conceded to reducing their proposal to a ten-cent cut, which was grudgingly accepted by the miners. However, the syndicate demanded another ten-cent cut almost immediately. Of course, the demand was refused by the union negotiators. In response to the union's refusal to accept the cut, the syndicate put out a notice that as of June 23, 1884, the pay rate would be sixty cents per ton. The only options were for workers to accept the wages the owners offered, look for other work or strike. The miners decided to strike.

Thus began one of the most well-known strikes in Ohio, in which miners held out for nine months against mine owners.[5] When the call for a strike was made, the syndicate responded by lowering the previous pay cut demand to fifty cents (another ten-cent cut) per ton and demanded that workers sign ironclad agreements to disavow their union and future union activity, further angering the miners. Forty-six mines stopped production, and more than three thousand miners were out of work. During the strike, families struggled to survive, sometimes living on what food union leaders could supply or what the miners could obtain by hunting and foraging. Violence was commonplace. The syndicate hired notoriously brutal Pinkerton guards to protect their interests. Syndicate agents went into southern states and hired unsuspecting Black laborers to take the place of striking workers. Foreigners were also encouraged to come to the mines, not knowing they were being used as strikebreakers. The miners reacted by meeting train cars in Nelsonville and threatening the people who were coming to take their jobs. The violence escalated. At one point, a coal car at the Plummer Hill Mine (between New Straitsville and Shawnee) was filled with burning objects and pushed into the mine, creating a coal fire that burns to this day. Owners had the wealth and resources to wait out the starving miners who were desperately trying to feed their families with what little they could get from an underfunded union and dwindling monetary contributions from the public. When it was all said and done, miners were forced to accept the rate of fifty cents per ton, with the exception of the New Straitsville miners, who had to accept forty cents per ton as punishment for the union leaders who lived in that town and led the strike. Months later, after litigation, a judge ordered the companies to return to the rate of sixty cents per ton for the Hocking Valley miners. However, the point had been made that the owners would do everything in their power to use workers as they saw fit. On the other hand, workers learned that without a more powerful union, they would never be able to achieve fairness or safety.

Slowly, miners managed to amass some power under the United Mine Workers (UMW) union formed on January 25, 1890, in Columbus, Ohio; however, even that powerful union was not always enough to get the miners what they were asking for. Many strikes occurred after the UMW was founded. For example, several years prior to the Millfield mine disaster (1927–

29) workers, under the leadership of UMW president John L. Lewis, went on strike and were met with formidable force by the mine companies, only to learn that not much had changed in the determination of mine owners to defeat unionization. An article from the *Canton Daily News*, published in 1928, provides an overview of typical tactics used, typical arguments coming from both workers and owners or operators and typical threats of almost all strikes since the early days of unionization.

> *CLIMAX TO COME FIRST OF APRIL IN MINE STRIKE*[6]
> *No Settlement Seen: Union Men to Be Evicted*
>
> *COLUMBUS, Feb. 13—Ohio's greatest industry—that of bituminous coal mining—was virtually paralyzed today with complete abandonment of many of the mines contemplated.*
>
> *Forty thousand striking miners are idle, their children are ragged, starvation is imminent, sabotage and rioting are frequent, and the climax is still in the offing.*
>
> *The strike—one of the most bitter dramas in the history of the country—has been in effect since last April when the Jacksonville wage pact guaranteeing skilled miners $7.50 a day expired.*
>
> *The pact possibly will never be renewed because the mine owners insist they cannot operate under the agreement and profitably compute* [sic] *with nonunion mines in Kentucky and West Virginia.*
>
> *The miners—led by John L. Lewis, president of the United Mine Workers of America—are determined to hold out for renewal of the pact or at least what they term an equitable compromise.*
>
> *And now, warnings have been sounded in congress and state militiamen are held in readiness prepared for eventualities.*
>
> *An ominous cloud of unrest, fraught with premonitions of disaster, hangs over the strike-torn coal fields and state officials are watchfully waiting.*
>
> *Open rebellion—if such occurs—may be expected April 1, officials fear, when a federal court evicting union miner's families from houses owned by non-union mines becomes effective.*
>
> *Meanwhile the state has leaped into the fray.*
>
> *A relief proclamation was issued by Gov. Donahey, and the entire state has responded. Sixty-five kitchens have been set up in the coal fields and are feeding 5,000 children a day. Scores of trucks loaded with clothing, shoes and canned goods have been sent to the destitute areas and the state has an $18,000 relief fund on hand.*

But over it all hangs the feeling of hate and unrest that has flared forth in rioting, arson and, in a few instances, murder.

Abject squalor prevails in many communities. The miners are sullen and subject to coercion, their wives are ragged, their children are in tatters and there is much sickness. The state's relief kitchens have been set up in the schools where the hungry children are fed, and this fact alone is believed to have staved off armed warfare.

Meanwhile the state's coal production has declined from 3,000,000 to 200,000 tons a month. Two hundred and fifty large mines—normally employing 30,000 men—are idle and 90 per cent of the smaller mines are in operation. In addition, the Sunday Creek Coal Co., one of the largest in the state, is threatening to abandon its mines, even to withdrawing its maintenance men.

W.W. Titus [sic], *president of the company, said that unless operations are resumed, the company will be denied its annual shipping contracts and it will be forced to shut down entirely. Other companies probably would follow the Sunday Creek concern's lead, Titus* [sic] *declared.*

By the time the strike ended, the former Poston Mine no. 6, which had long been a unionized operation, had become Sunday Creek Coal Company Mine no. 6 and was a nonunion operation, and workers were being paid what the owners/operators wanted to pay them.

Most certainly, there was at least some lingering resentment between workers and Sunday Creek Coal Company's leadership at the time of the disaster, which came only a year after the defeat of the union's goals. To what extent that resentment might have contributed directly to the company's inability or lack of desire to provide adequate safety for the workers is likely never to be known, but safety concerns continued as the company put men into the mines with outdated equipment and unsafe working conditions. Directly related to the disaster was evidence that either proper mine safety checks were not done, or obvious dangers were ignored. As will be seen in the investigations that followed the disaster, the company could have and should have done more to protect those who ultimately died—including the company's own executives and guests who were in the mine that day.

A BRIEF HISTORY OF SUNDAY CREEK COAL COMPANY, MINE NO. 6

Clinton L. Poston, while living in Nelsonville, Ohio, was actively involved in his father's businesses in Nelsonville, which included coal mining. After his father's death, Clinton expanded operations and became wealthy. He was a smart and successful businessman and coal producer. His leadership skills and keen eye for investments earned him the reputation of being one of the pioneers in the coal production industry in Ohio. One of his greatest skills was leveraging the wealth of his friends through partnerships to fund rapid development of infrastructure for each mine and hastening huge production capabilities from which he and his partners then sold at significant profit or leased to others to operate. These operators would then be obliged to provide a percentage of their profits to the parent company. Poston developed numerous companies, always gaining financially from them. In the late 1800s and onward, he began buying land and mineral rights throughout Dover Township, Athens County, knowing full well that there were huge profits to be made in extracting coal from these lands.* It was only a matter of time before his plan could be implemented.

Poston and others like him seemingly arranged and rearranged alliances and named and renamed or sold and leased properties throughout their lives, thus making it difficult to follow the complexities of ownership. For example, Poston and his brothers formed Poston Brothers to take control of their father's properties and ventures after his death. Clinton's brothers (William W., Ervin and Clarence) decided to sell their shares to McClintick and Smith, lawyers and businessmen of Chillicothe, Ohio, after which the business was known as C.L. Poston and Company. Eventually, Poston bought out his Chillicothe partners. He also bought and developed properties for mining through a company known as Poston-Fluhart. He brought all of his coal properties in the Millfield area together through the Millfield Coal and Mining Company and then formed the Poston Consolidated Coal Company, which would be responsible for mine no. 6, soon to be situated on the Dover Township land he had been purchasing. Meanwhile, he was leasing properties out to others, including the Sunday Creek Coal Company, and collecting royalties on the coal produced. The web of interconnections is vast and sometimes mind-boggling.

* This was the land where the Bailey Run Mine and Poston Mine no. 6 would eventually be.

Clinton, joining with other entrepreneurs, George K. Smith, T.R. Biddle, C.H. Horn and Lorenzo Dow Poston (Clinton's son), formed the Millfield Coal and Mining Company on December 3, 1910. Its articles of incorporation stated that it was

> *formed for the purpose of mining coal and dealing in coal, coke and kindred products; with power and authority to purchase, sell or lease mineral lands, and to purchase, own, lease or control such other property or real estate as may be suitable or necessary for the transaction of its business, including the right to construct, maintain and operate such spur tracks and sidings as may, for the transaction of its business, be suitable or necessary to connect with the main line of any railroad.*

The capital stock of the company at the time of incorporation was $50,000 divided into five hundred shares at $100 each. Converted to 2023 currency, the capital stock would be equivalent to $1,596,647.37. Divided into five hundred shares, the cost per share would be $3,193.29.*

Poston Consolidated Coal Company was established in January 1911 and took control of the Millfield properties where mine no. 6 was to be built and where mine no. 7 was being developed under the control of the Poston-Fluhart Coal Company. In the 1912 *Annual Chief Mine Inspector's Report to the Governor*, Poston no. 6 is described as a shaft mine situated in Millfield, Ohio, on the Kanawha and Michigan Railroad.† The shaft was 186 feet deep and provided access to Middle A no. 6 bituminous coal; the thickness of the coal bed varied between 4 feet and 6.5 feet throughout the mine.‡ At the time of the 1912 *Annual Report*, mine no. 6 was operating with a temporary tipple until a permanent tipple could be built; the mine had a double- and triple-entry system and fan ventilation, and the mine was equipped with electric power for machine mining.

The Sunday Creek Coal Company, which would take control of mine no. 6 in 1929, had been formed in New Jersey in 1905, prior to relocating to Ohio, where it was registered as a corporation on April 17, 1919. The people named as incorporators were John S. Jones, Chester G. Cook, George K. Smith, Harry B. Arnold and William Burry. Many of these names appear on other business incorporations related to coal mining—some of which

* Values determined through use of the CPI Inflation Calculator.
† At the time of the mine disaster, the railroad was owned by the New York Central Railroad.
‡ Sources differ on the depth. For the purposes of this book, the 1912 Mine Inspector's Report of 186 feet is used.

also included C.L. Poston.* The tangled web of interconnecting businesses and businessmen becomes more convoluted when research shows that eight months after Poston Mine no. 6 was under control of Sunday Creek Coal Company, a supplemental lease for more land and mineral rights for Millfield Coal and Mining Company was made to Poston Consolidated Coal Company. The lease, which does not mention the Sunday Creek Coal Company, was signed by "W.E. Tytus, President," for an agreement between Millfield Coal and Poston Consolidated. William E. Tytus was the president of the Sunday Creek Coal Company and one of the men who died in the mine explosion. Why he was signing on behalf of Poston Consolidated or the Millfield Coal and Mining company is unknown.

CONTEXT FOR THE DISASTER

The Millfield mine disaster story is first and foremost about the eighty-two men who died. However, as described earlier, it is also about many other issues. It is the story of a tragic event within a particular time in our history, when people who were trying to care for their families had to accept limited work opportunities in dangerous situations for low pay. It is also about an event within the convoluted relationships between wealthy entrepreneurs and stockholders—all of whom remained far removed from the lives of those who actually had to do the backbreaking and dangerous work that increased their wealth. And it is the story of ongoing issues in the United States that underlie our thoughts about such things as class inequality, corporate rights and privileges versus human rights and privileges and the issue of profit over people.

* Interestingly, George K. Smith was one of the incorporators of the Millfield Coal and Mining Company as well and took control of the Sunday Creek Coal Company after the Millfield disaster. Clinton L. Poston served on the company's board of directors at the time of its founding in 1905 and had business relationships with the incorporators.

2

THE MINE SITE

William E. Tytus became the president of the Sunday Creek Coal Company in June 1927 after serving in other positions under the leadership of then-president John Sutphin Jones, his father-in-law. As president, Tytus oversaw the operations of a number of mines around Ohio and beyond. Millfield no. 6 was one of the company's most recent acquisitions. Tytus had been engaged for much of the previous year in getting the mine up and running as quickly as possible after its purchase from Poston Consolidated Coal Company. Interestingly, on Thursday, February 6, 1930, approximately a month into operations under Sunday Creek, the record for the largest tonnage of coal in the mine's history was surpassed: 1,701 tons were produced in a single day.[7] To the company's credit, while a focus was on making profit, minimal efforts were being made to improve the most flagrant safety hazard: poor ventilation. This problem required digging a new airshaft and installing a powerful air fan; this project had not been completed by the time of the mine explosion. A new substation was also required and had just been completed.

Tytus and other officials and business acquaintances came to Millfield on November 5, 1930, to inspect the work done on the two improvement projects—and perhaps impress their guests or encourage them to contract with Sunday Creek. As the group made its way to the mine that day, they would not have been able to ignore the sights and sounds of the huge complex of buildings, coal cars sitting on train rails below the tipple, men working at various tasks, the powerful engines of the hoist house raising and

Top: Sunday Creek Mine no. 6 (expanse) from the W.E. Peters Collection. *Courtesy of the Mahn Center for Archives and Special Collections, Ohio University Libraries.*

Bottom: Sunday Creek Mine no. 6 in operation (looking southwest) from the W.E. Peters Collection. *Courtesy of the Mahn Center for Archives and Special Collections, Ohio University Libraries.*

lowering the elevator through the 186-foot-deep shaft, the hammer against the anvil of the blacksmith shop and coal banging and crunching as it was dumped from the hauling car in the tipple and then again as it came down into the waiting rail cars. Above them, a white and gray cloud of smoke filled the sky above the power station's smokestack. Little did Tytus and the men who accompanied him know that as the hoist lowered them down the shaft, what they had just seen and heard were to be the last they would experience in the light of day.

On November 5, 1930, the Sunday Creek Coal Company's Mine no. 6 consisted of numerous buildings shown on page 39 and identified in an architectural rendering that follows. Descriptions of buildings are taken from a report submitted to the National Park Service, U.S. Department of the Interior (Erickson, et al.) and a National Register of Historic Places application (1978):

a. **power station:** A steam boiler room and blacksmith shop, approximately fifty feet long and twenty feet wide.
b. **smokestack:** At the back of the power station was the smokestack—one of the remnants from the site still standing at the time of this publication, though it is in a state of significant decay. It is a rectangular chimney approximately 7 by 10 feet at the base and 150 feet tall.
c. **hoist house:** The building housed the hoist engines that raised and lowered the elevator in the mine shaft. Its original size is not known.
d. **mine shaft:** Measured 25 feet long by 14 feet wide by approximately 186 feet deep. It consisted of three compartments: a small compartment for carrying water and power lines into the mine and the other two compartments for hoisting.
e. **tipple:** The building, constructed of heavy hardwood timbers and steel bolts, was between fifty and sixty feet tall and was approximately sixty by twenty feet at its base.
f. **coal cleaning house:** It was used for the cleaning and grading of coal and feeding the coal product into four separate rail car lines.
g. **office:** There is no specific information available for calculating the size of the building or determining the materials used in its creation or how it was used.
h. **hospital***
i. **shower room and shower addition:** This was a one-story brick and tile building with a cement floor.

* This building did not exist in 1930. It was added in 1932 and was referred to as "the hospital."

Sunday Creek Mine no. 6 main shaft landing (view 1). *U.S. Bureau of Mines 40626; courtesy of the Technical Information Center and Safety Academy, 2011.*

Sunday Creek Mine no. 6 main shaft landing (view 2). *U.S. Bureau of Mines 40621; courtesy of the Technical Information Center and Safety Academy, 2011.*

Sunday Creek Mine no. 6 tipple. *Ohio Bureau of Mines 40624; courtesy of the Technical Information Center and Safety Academy, 2011.*

j. **airshaft and fan house:** The airshaft is lined with concrete subterranean walls. The fan house measures approximately twenty feet long by fourteen feet wide. A seven-foot fan pushed air but was also reversible.[8] This fan normally ran on "six points," or 75 percent of its full speed; it was equipped with a sixty-inch pulley belted to a twenty-inch pulley on a variable speed induction motor.[9]

k. **tram house:** This was a small structure on the north side of the tipple. It was used for moving a tram bucket along a cable to the gob pile.

l. **gob pile:** A place to pile the waste removed from the coal during the sorting process.

m. **tram bucket cable:** This was used to carry waste materials from the tram house to the gob pile.

n. **loafing shed:** A wood-frame structure used by the men who were waiting to go down into the mine shaft via the elevator.

o. **water tank:** A circular, double-walled brick structure reinforced with steel rods. Located on the north side of County Road 27. It supplied water for the shower room.[10]

p. **sand house:** A one-story unglazed brick building that measured ten by twenty feet. The sand house was a small brick building that was used for storing sand shipped from the Great Lakes to be used in the mine when cars slipped on the track.[11]

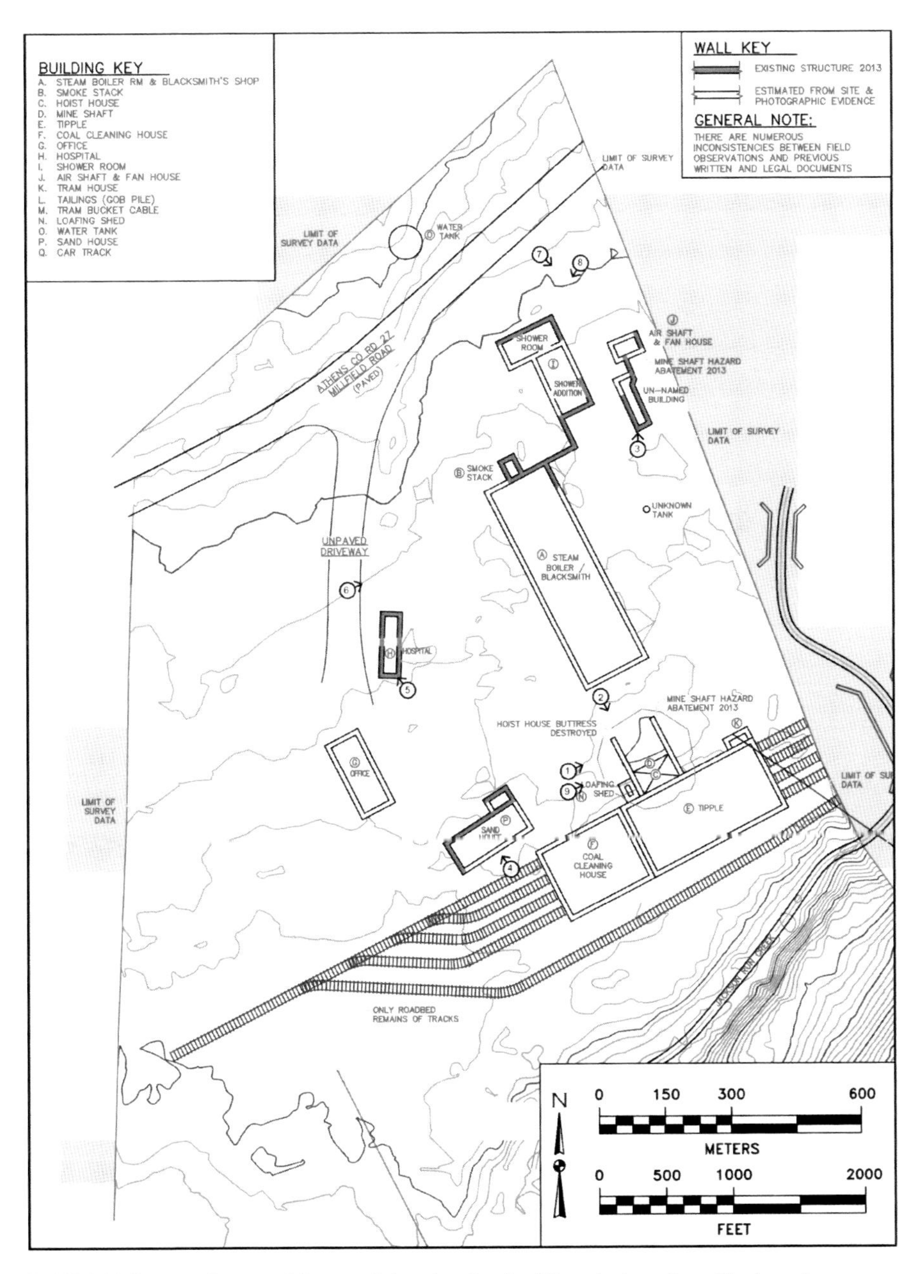

Millfield Mine no. 6: an architectural drawing for the Historic American Engineering Record, National Park Service, U.S. Department of the Interior, HAER OH-139. *Courtesy of David Reiser, Architect, Athens, Ohio.*

Traces of other buildings appear on the property and can be seen in extant pictures, but their purposes cannot be identified.

Not shown in the architectural rendering on page 39 is the new airshaft that had been created during the months after Sunday Creek took ownership from Poston Consolidated Coal Company. Directly above the 176-foot-deep shaft, a new fan was being installed, and a structure was being built to protect it. However, the fan's installation was not far enough along for it to be put into use at the time of the explosion. A straight line between the main entry shaft and the new airshaft would measure approximately 7,100 feet (or 1.34 miles).

Also not shown in the architectural rendering is the company store, which still sits in the town of Millfield at the corner of Main Street (County Road 93) and Millfield Road. The Sunday Creek Coal Company purchased the Columbian Store Company's building in January 1930 and opened it* to the miners in early February.[12] At the time of the disaster, it was used as a temporary morgue.

Company president Tytus and his cohorts apparently were unaware of or were unconcerned about any significant danger as they entered the elevator/hoist and descended to the bottom of the shaft. It is highly unlikely that they would have entered if they had thought that not all precautions had been taken prior to their arrival. It is impossible to know, of course, what they might have been thinking, saying or worrying about as they stepped into the main passage that led to the mine's more complex grid of passageways and honeycomb-like "rooms" carved into the coal deposits.

Mine no. 6 used "room and pillar" mining. This method was commonplace in bituminous coal mining. The method involved creating a grid of intersecting passageways through which miners could reach work locations and coal cars could be towed to where workers could most efficiently fill them before sending the coal back to the main shaft to be raised to the tipple for processing. The mine consisted of parallel and perpendicular tunnels that extended at least a mile and a half into the coal beds that had been started under Poston Consolidated and had expanded during the brief time of Sunday Creek's control. The layout was much like that of a city. Tunnels gave access to "rooms" like streets give access to buildings. The rooms' sizes and shapes depended on the coal's accessibility and its projected value compared to the costs required to retrieve it.

---

* George C. Duncan of Murray City served as the person in charge. The building was the same building that the Poston Consolidated Coal Company had used previously.

Sunday Creek Mine no. 6: Fan being installed on new air shaft. *U.S. Bureau of Mines 40627; courtesy of the Technical Information Center and Safety Academy, 2011.*

Sunday Creek Mine no. 6: a fan motor and transformers for new air shaft. *U.S. Bureau of Mines 40628; courtesy of the Technical Information Center and Safety Academy, 2011.*

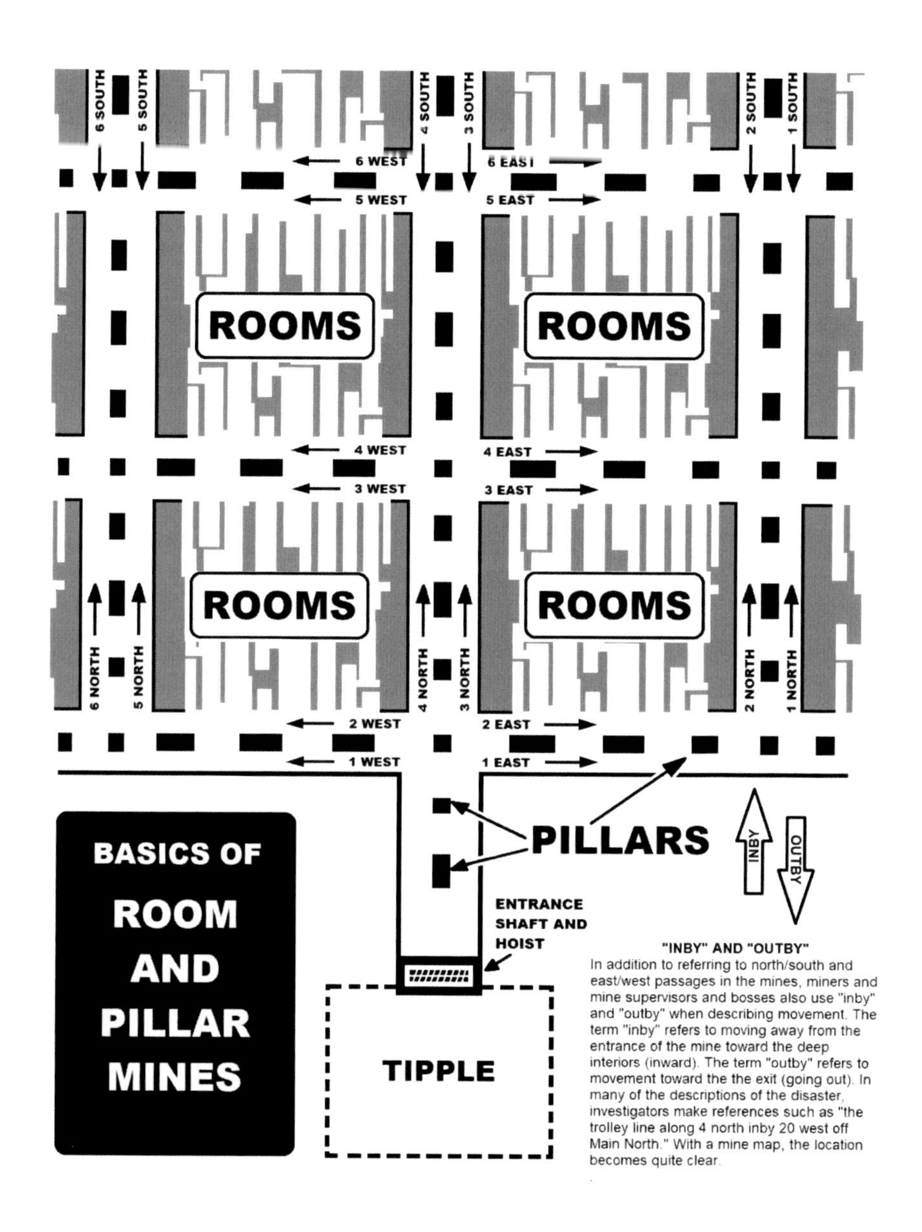

In the best of circumstances, just enough coal was left in place in these rooms and tunnels to support the weight of the mine "roof" so that maximum extraction could be achieved. This method of room and pillar mining allowed coal companies to extend miles underground until all available coal could be removed.

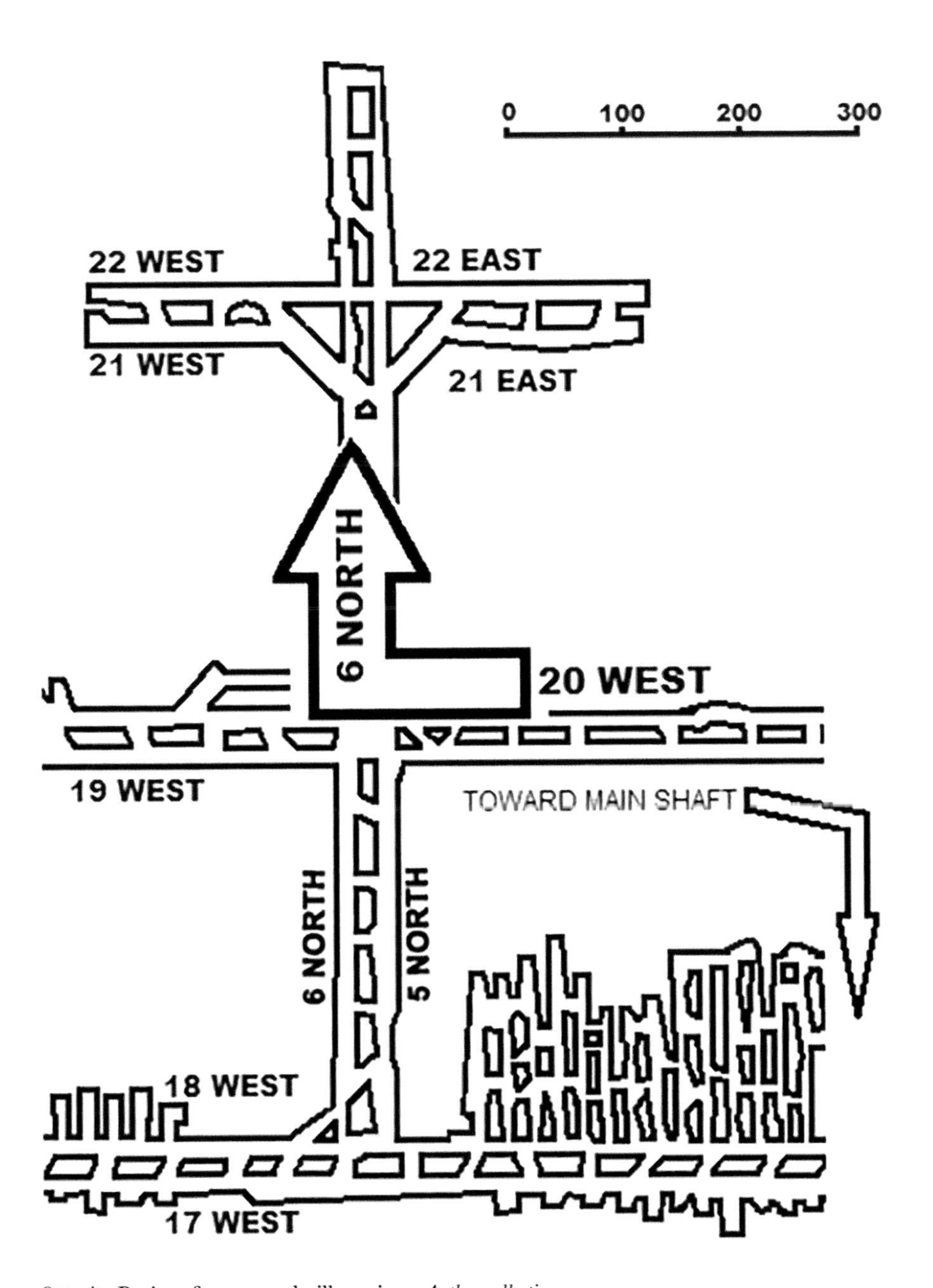

Opposite: Basics of room and pillar mines. *Author collection.*

Above: Mine directions example. *Author collection.*

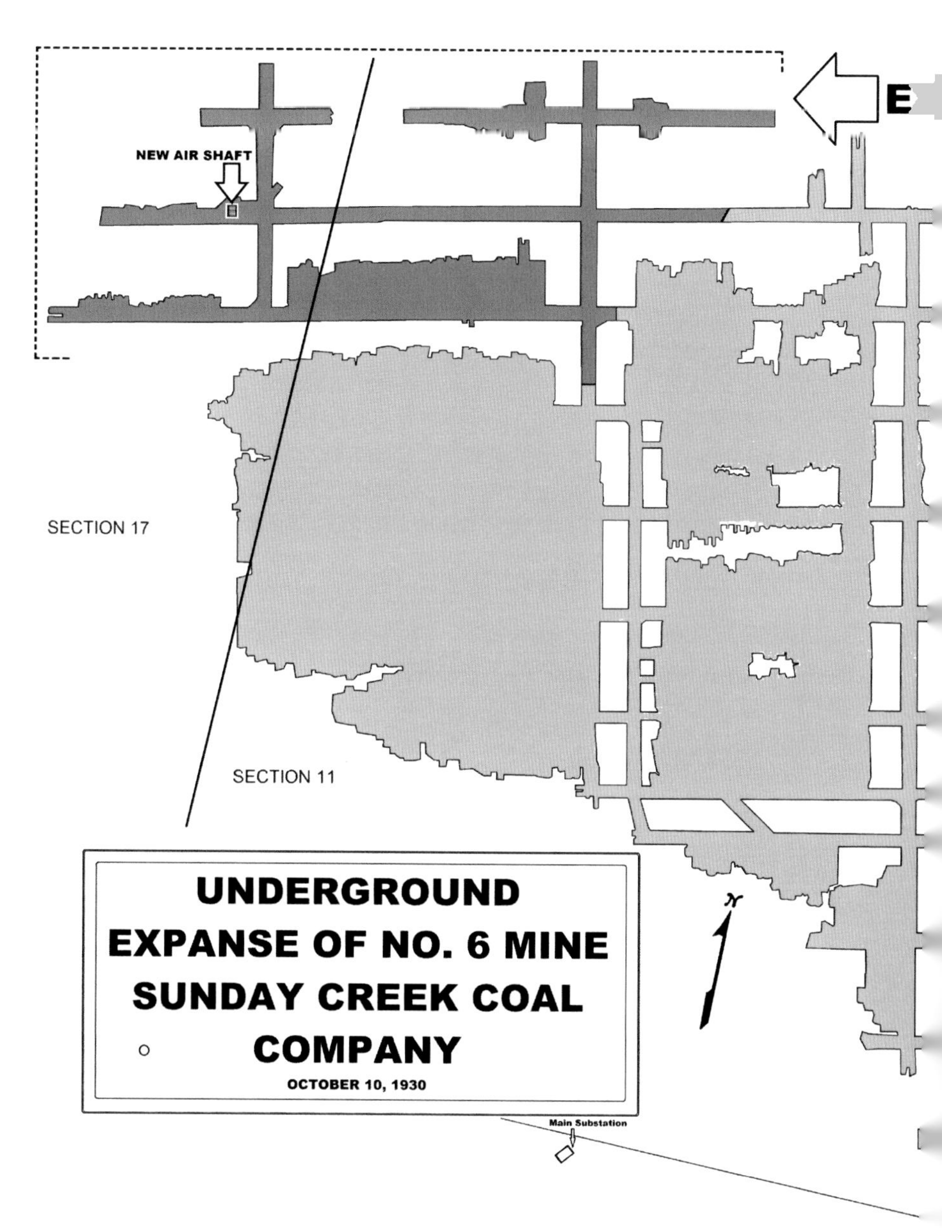

Athens County, Ohio: Range 14, Town 10

Situated primarily in Section 11, but extends into Sections 5 and 17.
At the time this map was drawn, the underground works extended slightly over a mile (north and south) and 1 1/2 miles (east and west),

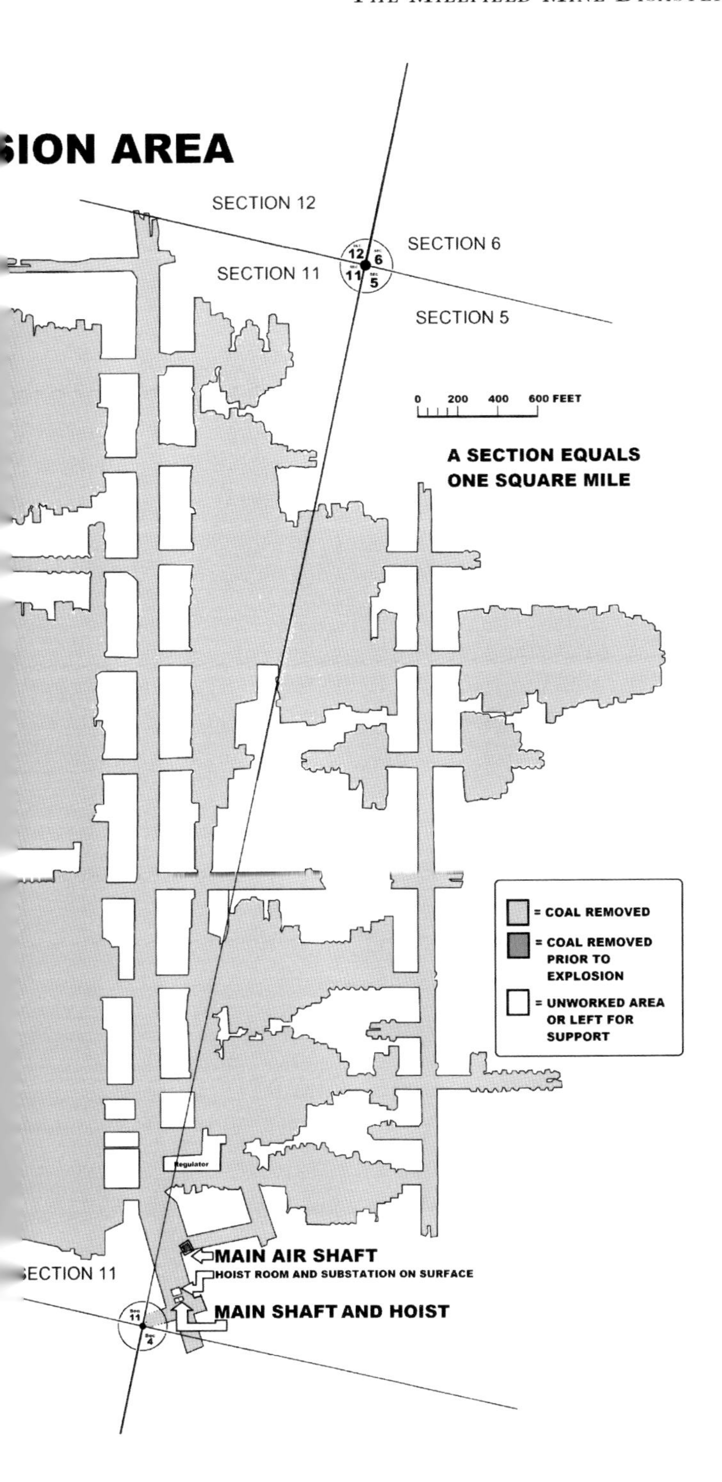

Underground expanse of no. 6 mine. *Rendering based on a map from the Ohio Department of Natural Resources (ODNR) website: https://ohiodnr.gov. Author collection.*

Miners use terms for the directions in a mine that can be somewhat confusing to people who are unfamiliar with them. It is helpful to the reader who may be looking at the illustrations and trying to locate specific points in mine no. 6 to become familiar with some of the terms miners used to explain directions. All tunnels are named and numbered based on whether they run east–west or north–south. When making a turn at any corner within the gridwork of tunnels, a miner is going either "inby" or "outby." If the turn takes the miner away from the main shaft and farther into the mine, the miner is moving inby—going further *into* the mine. If the miner takes a turn that moves in the direction of the main shaft, the miner is going outby—going out of the mine. These terms help pinpoint where someone or something is. In the case of the Millfield mine disaster, for example, a body was found "on 6 north, 300 feet inby 20 west." Though it sounds confusing at first, it provides a very specific place to look: the body is in tunnel 6 north; to get to that body, a person would go north on any north–south tunnel to 20 west. Taking 20 west, the person would go to the intersection with 6 north and take 6 north for three hundred feet to where the body would be found. It is very much like saying, "Go north until you come to Mulberry Street, take a left on Pacific, which runs east and west. When you get to the intersection with Elm Street, turn right on Elm. Shortly after you make the turn, your destination will be on the right."

Each tunnel is further divided into two distinct paths separated by columns. Often, brattices and doors were installed between these columns and elsewhere to maintain airflow control between incoming fresh air and return air. Except for portions of the two main haulage routes—8 west and a portion of 4 north off 8 west—rail was placed on the return air side of the tunnel. These rails were used as part of an electric trolley system that could move cars to desired locations for loading and unloading coal. The rails provided haulage on forty-two-inch-gauge forty-pound rails.* In general, five and a half feet of space was left on both sides of the track. The tunnels' sides were divided by somewhat evenly spaced pillars of coal that were left in place to support the "roof" (slate) of the mine. At the time of the disaster, this vast network of rooms, pillars and passageways took up at least one and a half square miles of space underground.

* The gauge is the distance in inches between parallel rails; "forty-pound" refers to the weight of the rail at forty pounds per yard.

3

EXPLOSION

Wednesday, November 5, 1930

All of the men making their way to the mine that morning carried a load of responsibility on their shoulders. Most had families and were living during the Great Depression, when the unemployment rate was well over 18 percent, jobs were scarce and pay was low. Though pay had been cut significantly over the previous few years and little had changed in the way of making the work safer, it was considered better to have a job than to risk putting themselves and their families in a state of destitution. It is likely that most of the men had never known anything other than coal mining; many had learned the trade from their fathers, grandfathers, uncles, brothers and other miners. Many had started going into the mines with their fathers to help increase the father's output; when they were capable of working on their own, they started taking whatever paying work they could to supplement their families' meager incomes. Because coal mining families prior to 1930 often moved from coal camp to coal camp in search of better pay and steady work, few of their children gained a full elementary and high school education and likely had few opportunities to learn other trades. Their identities were shaped by the people they lived with in both their homes and their communities. Generations of miners clung to their identities as miners regardless of whether it was in their best interests to do so or not. Many seemed to accept that there were many factors beyond their control that dictated whether or not they could make enough money to survive the Great Depression.

Miners had been defeated by the companies they worked for and devastated by the Depression. As a result, they simply dealt with circumstances as best they could. When it was all said and done, they took what work they could find at the price the companies were willing to pay, swallowed their pride and made do while they dreamed of better days in some far-off future. Like most people of that era, the miners had little choice but to focus on things other than the ugliness of their existence: they had families, friends, lovers, ambitions, talents, dreams and desires. They focused on the people they loved, the pleasures they could find and the pride of being good at what they did as workers—whether it was appreciated or not.

On the morning of November 5, miners and their families rose to what probably seemed a typical early November day. There was a chill in the air. Throughout the morning, the temperature would rise to a high of 51 degrees Fahrenheit at around noon. The barometer showed that air pressure had fallen slightly from the previous day, and that trend would continue, reaching 29.45 at noon—a standard air pressure.[13] The men, their wives and families got out of their beds and, perhaps, ate breakfast of one kind or another. Parents nudged their younger children to get their schoolbooks together, checked to see if they had properly washed and were dressed appropriately. Then they all said their goodbyes and went their separate ways.

The men headed to the mine to carve the "black diamonds" out of ever-expanding mine tunnels. Though all miners knew there were dangers in the work they did, most did not live in fear of it. Most likely, they took a fatalistic view of it: if it happens, it happens; a man's got to earn a living; best not to think about it too much; et cetera. Undoubtedly, routine kept them focused on going to work; coming home later in the day; maybe doing a little gardening work, visiting neighbors or playing with their children; working on some household project or any number of other human activities that make a life. Women with preschool-aged children tended to them amid cleaning their homes, preparing meals, washing clothes and performing the other endless tasks of motherhood and housekeeping. Perhaps some of the women took time out to go to the company store for food and maybe look at the items they could not afford and took whatever pleasure they could find in dreaming of owning "nice things" as they returned to their homes to prepare dinner for their families.

Most workers at mine no. 6 came to work from within the immediate area of Millfield, but some made their way over the hills from Canaanville, Glouster, Jacksonville, Guysville and other nearby communities.

Millfield, Ohio street and store, pre-1930. *Author collection.*

Millfield itself was a small town, but it had two distinct sections. Many of the miners—particularly the foreign-born miners—lived in what is referred to as "East Millfield," or the "other side of Sunday Creek," the side of the town where the poorer people of the community lived and where miners were housed in company-owned buildings.

On the morning of November 5, miner Oscar Willis said goodbye to his wife, Minnie, and his younger children—his son, Gerald, and daughter, Kathleen—both of whom would soon be heading off to school. He also said goodbye to his son Charles, who would ordinarily be going with him to the mine but who was ill and couldn't work that day. Charles, like his father and two older brothers—Andrew and Virgil—was a miner and would ordinarily be jostling with his brothers to grab his lunch pail and gear to get out the door as the men made their way to the mine on time. No one knows for certain what might have been on their minds that day, but it is not too difficult to imagine that perhaps Andrew (who played a mandolin) and Virgil (who played a violin) were thinking about the next opportunity for their band to get together. Or perhaps the young men were thinking about riding around in their convertible and trying to impress the young ladies of the town. Maybe they were just thinking about the day, things they wanted to talk about with coworkers, things they had to do after work. Oscar might have been talking with his sons about some family issue.

Above: East Millfield company-owned housing for miners. *Author collection.*

Right: Oscar and Minnie (Bolton) Willis. *Courtesy of Jim Mingus.*

Opposite, top: Members of the Willis family (*left to right*): Andrew, Oscar, Charles and Gerald (*front*). *Courtesy of Jim Mingus.*

Opposite, bottom: The Band (*left to right*): Elmer Davis, Charlie Simpson, Virgil Willis and Andrew Willis. (Behind the band and between Virgil and Andrew is Charles Willis.) *Courtesy of Jim Mingus.*

Alfred Wade and his son Luther might have spent their time that morning expressing their perceptions of their first two days on the job or the complications of the move they had just made from Canaanville to Millfield.[14] John Williams may have been concerned about his wife, who was ready to give birth to their fifth child at any moment while managing the burden of caring for four others under the age of seven. One can only guess at the responsibility Ellsworth McKee must have felt and thought about as he walked out the door of his home that morning, leaving a wife he had married in March, their three-month-old child and a mother and sister who lived with him—people who depended on his income for their survival. Seventeen-year-old Andy Kish Jr. might have been thinking about any number of things—the stretch of lifetime in front of him and how he would spend it, young women he knew who caused him to think of romance, the changing of his body from boy to man or any number of things a boy of his age might concern himself with. No one will ever know. No one will ever know what was on the minds of any of the eighty-two men making their way to the mine that day or at the moments of their deaths a little before noon. What can be known is that they were men who had lives to live, people to love, families to care for and people who loved them, depended on them and would mourn their loss.

The day had started earlier for two fire bosses—Benjamin Fielder and Phillip Powell—who were required to be in the mine before the other miners. Their duties were to do pre-shift inspections of active works, checking for gases and other potential hazards. This work was to be done between 3:00 a.m. and 6:00 a.m. and again after 6:00 p.m. Additionally, they were expected to check inactive portions of the mine two times a week.[15] How they went about this task and who specifically was responsible for overseeing their work is not clear. As it later came to light, they were often told to spend their time doing "more important" things, such as building brattices.[16] Given that the company's president and other officials were coming to visit, it seems odd that a thorough assessment of safety was not ordered; if it was, it was either neglected or done haphazardly, or the fire bosses were pulled from the task. Since both fire bosses died in the explosion, it will never be known why it was not done.

Company officials and guests who had come for an inspection were as follows:

William E. Tytus, president, Sunday Creek Coal Company
Pearl A. Coen, vice-president, Sunday Creek Coal Company
Howard Upson, assistant to President Tytus, Sunday Creek Coal Company
Hubert E. Lancaster, chief engineer, Sunday Creek Coal Company
Walter Hayden, superintendent, Glouster, Sunday Creek Coal Company
Thomas Harley, mine foreman
Joseph A. Bergin, superintendent, Ohio Power Company
Robert Day Parsons, vice-president and manager of the Columbia Chemical Division of the Pittsburg Plate Glass Company
Vernon Lyle Roberts, foreman, Columbia Cement Plant
Thomas Brush Trainer, traffic manager, Columbia Cement Company

Various articles suggest that the "guests" may have been potential coal buyers that Tytus wished to impress. Other articles suggest they were friends or business acquaintances who made their visit simply to spend time with Mr. Tytus. What time the visitors arrived is not specifically known, but most articles suggest they were in the mine for approximately half an hour prior to the explosion.

Approximately 225 employees were working underground on the day shift.[17] Until 11:45 a.m., the mine was operating normally. Cutting machines were loosening coal, coal was being shoveled into carts, carts were moving the coal toward the main shaft and the main shaft was carting full cars up to the tipple and lowering the empties for refill. The miners worked steadily to meet the expectations of the bosses for production. Their clothes and bodies became ever filthier from the coal dust that lay everywhere, lingered in the air and clung to their skin, their sweat causing it to be soaked into the fibers of their clothes and the fiber of their being. Those overseeing the miners during the visit were John Dean, mine foreman, and Robert Marshall, assistant mine foreman. The Tytus entourage had made their way to the new substation. It is not known whether the visitors had already seen the tunnel-level portion of the new airshaft or were planning to view it after their substation inspection.* It seems that all was going as planned and the mine was working normally when, suddenly, a circuit breaker at the hoist engine room near the tipple automatically tripped at 11:45 a.m. and all hell broke loose underground.[18]

The explosion began in a section of the mine referred to as 21 and 22 east off 3 and 4 North (see "Disaster Area Illustration," pages 54–55). A trolley wire was broken by a large piece of coal that fell from the roof of the mine. When it fell, the live electrical line arced on one of the steel rails used for hauling coal out of the section. When it arced, it set off an explosion of the methane gas that had accumulated in that portion of the mine, sending a flame outward, which then ignited explosive coal dust that was both in the air and lying about the mine. The accompanying flame scorched everything in its path. The explosion was powerful enough to throw men who were directly in its path against walls with enough force to kill them instantly. It threw heavy machines, such as tow motors, off their rails and tore up and bent steel I beams as it raced to the nearest release point: the new airshaft. According to Frank Ray, a consulting mining engineer who investigated causes of the explosion, a man by the name of B.H. Pellet reported his

* Since the airshaft was sealed at the tunnel level, it is probable the visitors intended to see the aboveground portion of the shaft after their inspection of the substation.

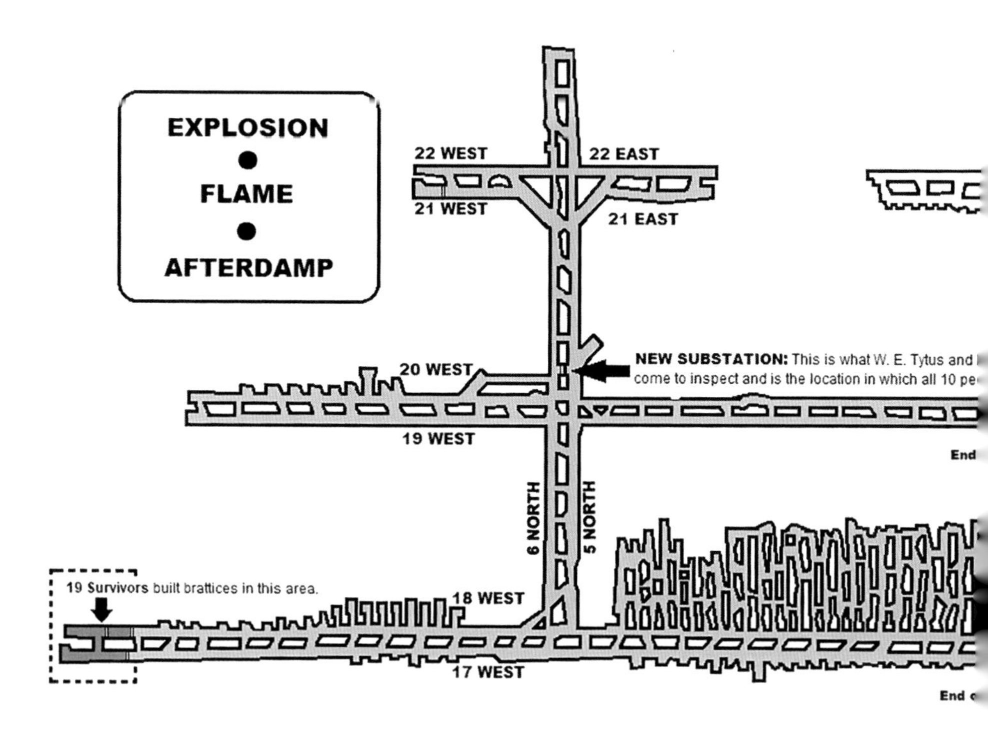

Disaster area illustration: explosion, flame and afterdamp. *Author collection.*

son had seen Ed Dempsey blown from the top of the new airshaft's fan equipment and land approximately fifteen feet away.*

A characteristic of these kinds of explosions in bituminous mines is that immediately after they happen, there is the sudden creation of "afterdamp"—a poisonous gas that, when breathed into the human body, combines with hemoglobin in the blood and deprives the victim of oxygen. As it turns out, this was the major cause of death among the eighty-two men in the mine: two men were killed by the force of the explosion, six by burns and afterdamp and seventy-four by afterdamp.[19]

At the time of the explosion, 119 men who were farther west of the explosion in the 7 and 8 east–west and 9 and 10 east–west sections of the mine—though thrown about—were not seriously injured and were able to escape.[20] They were led by assistant mine foreman Robert Marshall, who

* Mr. Pellet's son had been working nearby and was walking home at the time. Apparently, Mr. Dempsey had been working on top of the new fan.

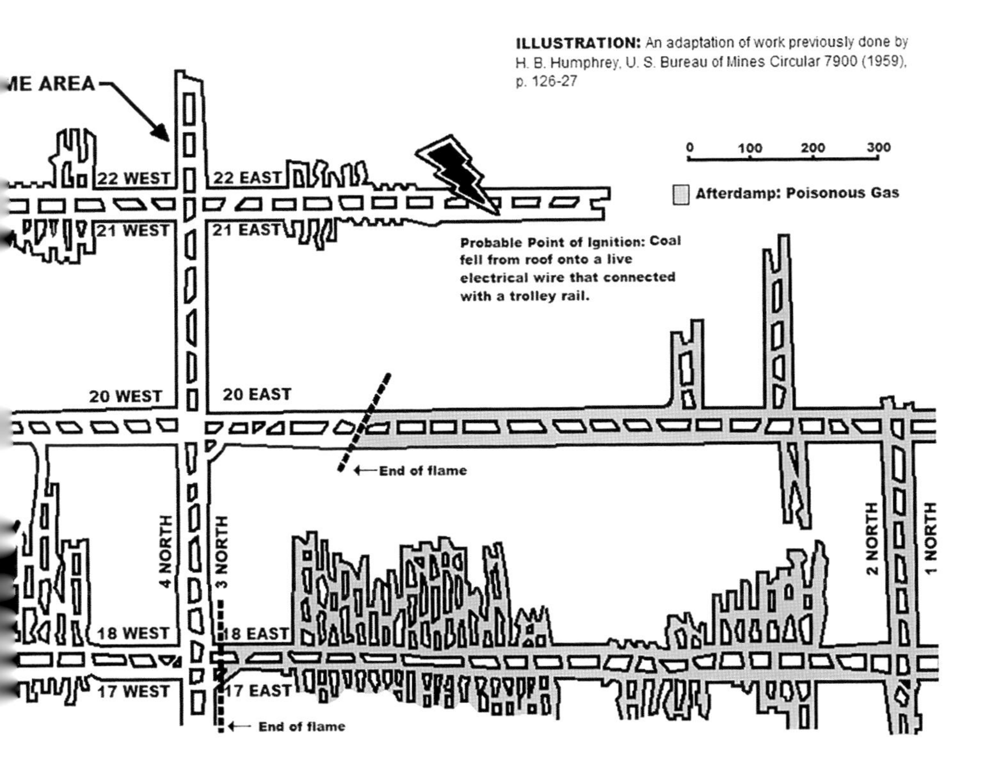

was nearby.[21] According to Walter H. Tomlinson,* Marshall is said to have "noticed a sudden rush of air and then a stoppage.[22] He gathered his men and led them to the main hoisting shaft. At that point, the main return contained little of the afterdamp."[23] By the time Marshall and others arrived at the main shaft, calls had already been made to obtain assistance in rescue and recovery.

Immediately after seeing Ed Dempsey blown from the new airshaft and confirming that he was not seriously injured, Pellet's son rushed home. He or a member of his household tried to call the local mine officials to notify someone of the incident. When there was no answer, a call was made to the Glouster office to notify people there.[24] Obviously, the Glouster officials responded immediately. Indicators of a life-and-death situation caused by an explosion were communicated to mine officials and the community. Calls were made to officials of the Sunday Creek Coal Company headquartered

* Walter H. Tomlinson was a foreman miner working for the United States Bureau of Mines.

Family members, volunteers and reporters gather at Sunday Creek Mine no. 6. At times, the road was so congested that rescue crews and public officials had to either walk in or wait for authorities to clear it. *Acme Newspictures, photograph, 1930; author collection.*

in Columbus, Ohio; E.W. Smith, the chief of the Division of Mines in Cambridge, Ohio; the Bureau of Mines in Washington; and numerous other entities that might play a role in rescue operations. A call was made for a rescue car to be brought as quickly as possible from Pittsburgh, as there was limited rescue equipment available locally.

No one knew for certain what would be needed, how many miners might be dead or how many might be hanging on for dear life. The miners who were out of the mine quickly reported the conditions underground and made clear that the mine was too dangerous for anyone without a respirator.* In an attempt to prepare, people took it upon themselves to ask for anything and everything that might help. Local communities were asked to send doctors, nurses and volunteers. Apparently, for quite some time, there was no individual to take charge, make decisions and coordinate rescue efforts. All of the leaders of mine no. 6 and the Sunday Creek Coal Company were

* Unfortunately, there were far too few respirators available, and more would have to be shipped in from elsewhere.

The crowd waiting behind a rope or cable. *Author collection.*

down below, and there was no one who could easily take charge and direct rescue activities until Robert Marshall found his way out of the mine and was soon joined by Andrew Ginnan, deputy mine inspector, who arrived at approximately 1:30 p.m. However, their focus was on rescuing anyone in the mine who might still be alive; neither had time for dealing with all of the decisions that needed to be made when it came to dealing with the problems aboveground, where concerned families and community members, medical personnel, reporters and well-meaning volunteers began rushing to the grounds, blocking the roadways and making it difficult for people needed for rescue work and caring for the injured or dead to get to the site. It became "necessary to call in police and the local national guard to handle traffic and the ever-expanding crowd. Every vacant street and lot in Millfield was used for parking and traffic was jammed. Officers were needed constantly to control traffic from Wednesday shortly after the explosion through Sunday when funerals were taking place."[25]

The text of the *Report of Gas and Dust Explosion in Mine No. 6, Sunday Creek Coal Company, Millfield, Ohio, November 5, 1930* (hereafter referred to as *Report*) from the Bureau of Mines provides significant information about what happened.[26] As early attempts to deal with the crisis were taking place, approximately an hour after the explosion, six men climbed out of what was referred to as the "manway compartment" of the new airshaft. George Rasp, an arc wall machine man who was the first man out, described the escape of the six men; his version of the event was confirmed later, in part, by Joe Reynolds and B.H. Pettit. Rasp said that he was followed by Frank Shumway, Lester Shumway, Joe Reynolds—who was working in a room off of 20 west at the time of the explosion—Steve Butsko and a miner named DeVore (his first name is unknown). DeVore was weak from the gases and needed help to get to the top. A carpenter, Ted Beal, who had been working aboveground on a project related to the new airshaft, went down two flights and assisted the man. A seventh man, Emerson Lefever, collapsed six flights down. James Mackey, who was a fire boss at the Ohio Collieries Mine no. 255 and off his shift at work, had come to volunteer and went down the airshaft to assist Lefever. He carried Lefever up two or three flights and then called for help, as he too was succumbing to the effects of the afterdamp. Again, Ted Beal went down the manway along with another unidentified man and carried Mackey and Lefever out. Lefever was given artificial respiration for approximately one and a half hours, but ultimately, he could not be resuscitated.

Elsewhere in the mine and after he had made sure his miners were safely getting out, Robert Marshall, accompanied by Walter Patterson and Mickey Wallace, both machine men; Virgil Rutter and Elmer Davis, both trip riders; and William Monroe, a trimmer, went back into the mine in an effort to save others.[27] They found Frank Williams, a machine man, near 16 west and 4 north.[28] At first, the rescuers thought Williams was dead, as he was badly burned. However, they carried him to an area where there was still fresh air at 13 west and 4 north and gave him artificial respiration for about an hour. When they could see that the man was breathing on his own, they improvised a stretcher and carried him out. According to the *Report*, "This was the first injured man to reach the bottom of the main shaft, where they arrived about 1:30 p.m."[29]

About the time Williams was being lifted out of the mine, Ohio's deputy mine inspector, Andrew Ginnan, arrived. Mine no. 6 was one of those that was within his jurisdiction. Under Ginnan's direction and personal involvement, Robert Marshall and others began the process of trying to direct fresh air into the mine so that rescue could be attempted. They went into the mine accompanied by Superintendent Peter H. McKinley; a section foreman; George Jackson, a machine man; and two motormen. The motormen advanced into the mine as far as the 13 and 14 west entries, where they cut power to the trolley wire. When that work was completed, a plan was created to clear the track of wrecked equipment so material and supplies for restoring ventilation could be brought into the mine.[30] In none of the available reports is this stated, but it would be logical that additional workers were brought in for the task of clearing debris and building brattices to divert fresh air out in front of them while the rescue crew moved deeper into the mine in search of potential survivors. As far as can be discerned, only one of the rescuers, Jake Maurer, had a self-contained oxygen breathing apparatus.* Thus, these early rescuers had no choice other than to proceed slowly toward the explosion area, making certain they didn't get caught by the dangerous gases. Much of their time was spent trying to move fresh air currents ahead of themselves. This meant dragging in building materials to make brattices and manipulating various configurations of brattices and stoppings.

As this first crew worked tirelessly at their rescue attempts, other professionals were doing all they could to get to the mine as quickly as possible. At approximately 4:00 p.m., Ohio state mine inspector E.W. Smith and deputy inspectors Isaac Vaughan, Thomas A. Richards and P.W. Moore arrived, bringing with them some equipment and testing devices, including canaries.† Apparently, they managed to meet Ginnan's team in the mine and joined the search. When they got to 15 and 16 west entries, the team

> *found all the stoppings burned, timbers displaced, cars and motors blown from the track and wrecked, and the tracks strewn with wreckage. We also*

* Whether Ginnan and others had gas masks of some sort on at that time is not known.

† Caged canaries were often used in mines to test for carbon monoxide in the air; the birds were susceptible to the toxic gas and would succumb sooner than humans would. When the canary died, it was an indicator that humans needed to get to fresh air as quickly as possible.

> *found the air at this point carrying a high percentage of carbon monoxide gas. A stopping was erected on 4th north entry inby, the 18th west entry and ventilation carried up the 17th west by canvass stoppings....By 6 o'clock, we had reached the 10th room on this entry where we found the body of the first victim of the explosion. It was that of Thomas Hardy* [Harley], *a safety foreman, who had been killed by violence. As we proceeded further along this entry, we found four more...badly burned and crushed by violence of the explosion.*[31]

Shortly thereafter, twenty-five more bodies were found. All of these men died from the effects of the carbon monoxide gas. They were all lying flat on their faces and as close to the floor as possible. Smith stated that everything indicated they had survived the force of the explosion "and had, no doubt, lived for some time after the explosion took place."[32]At that time, members of the rescue team assumed it was unlikely that anyone could have survived the catastrophe.

Chief Inspector Smith left the rescue crew to determine whether there was evidence of fire or smoke in the return air. Finding no evidence of a fire, he went to the main shaft and organized crews to carry on the work of clearing roads (tunnels and removing the dead. Since electricity couldn't be used for hauling due to the damage to rails, stock animals (horses, mules, donkeys, et cetera) were borrowed from nearby mines to assist in clearing tracks and removing bodies while an advance party continued to explore. Meanwhile, the rescue team, still being led by Ginnan and Marshall, was forced to stop when, at 17 and 18 west entries, the canaries that had been carried into the mine to provide warning about air quality were overcome. Again, the search was thwarted, and time was spent building brattices to move the air in the desired direction before the team could proceed.

Once the air was moved, the rescuers proceeded on to 5 and 6 north and eventually reached the intersection of 19 and 20 east–west. A short distance north of the intersection, they found the bodies of Tytus and the nine men who had accompanied him. The ten dead men had all been directly in the line of the explosion's violent force; some were both burned and physically mutilated. Smith asserted that, had any of these men somehow survived the blast, they could not have survived the effects of the afterdamp over the long distance to the main shaft.[33]

With his breathing apparatus, Jake Maurer was able to go a bit farther on 5 and 6 north, where he found two more bodies on what was referred to as "a sidetrack."[34] The team then returned to the intersection of 19 and 20

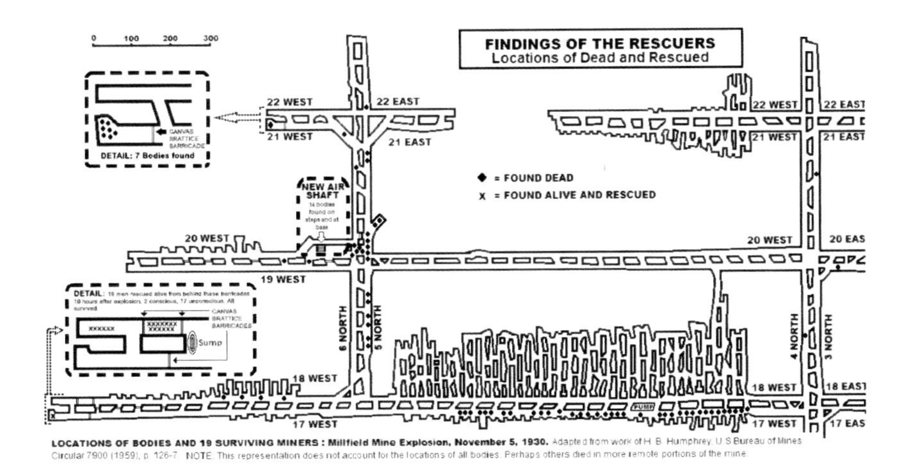

Findings of the rescuers: locations of the dead and survivors. *Author collection.*

west and reached the new airshaft at approximately 8:00 p.m.[35] The *Report* states that "by a calling up the airshaft to [Ohio] Chief Mine Inspector E.W. Smith, Ginnan informed him that men found were dead."[36]

It is not explained in the *Report* why Ginnan and other rescuers then returned to the point where they had begun building stoppings; it said on!- that they returned. Perhaps a messenger had come for them, or someone had yelled down the new airshaft to tell them the other deputy inspectors had arrived and that they needed to meet them. Whatever the reason, they returned and met Deputy Inspectors Elmer Sagle and Val E. Brown; the men had a brief consultation and then Sagle and Brown were escorted along 17 west with the crew building stoppings as they went.[37] Shortly before 9:00 p.m., the team was approximately six hundred feet from the face (the dead end of the tunnel) of 17 west. Sagle stopped. He had heard a whistle—the kind only a human being can make—and it was coming from behind a barricade that had been erected at the end of the tunnel. Fresh air was immediately conducted toward the area, and rescuers moved as rapidly as they could safely go to whoever might be there. Upon their arrival, they found nineteen men who had put up a barricade to protect themselves from the afterdamp. The survivors were carried out to an area of the mine where fresh air was being pumped. Artificial respiration was given to the men, and ammonia was used on some to bring them around as the rescuers waited for doctors who had been called in from aboveground. Eighteen of the men were wrapped in blankets and eventually carried out on stretchers; one man, Harold Phillips, was able to walk out on his own.[38] The eighteen miners who were carried out "were given oxygen intermittently for several hours before being removed to the hospital or their homes." The nineteen men were as follows:*

Ayers, Floyd
Cevelo, John
Channel, Ralph
Cobb, Robert
Crabtree, Floyd
Davis, Howard
Dean, John
Forsbach, Henry
Handa, James
Hunter, Earl
Hunter, James
Parker, Chris
Parker, Edward
Phillips, Harold
Pickering, Carl
Norton, James
Rinaldo, James
Watson, Clifford
Willis, Fielden

When arrangements could be made, fifteen of the survivors were taken to Sheltering Arms Hospital in Athens. Fielden Willis was taken to Cherrington Hospital in Logan. Three others, Robert Cobb, Floyd Crabtree and Harold Phillips (who had been able to walk out of the mine on his own), were taken to their homes.

Interestingly, a questioning of the survivors led to the surprise finding that the nineteen men had built the barricade four or five hours *after* the explosion. They were most likely able to hold out longer than others in the mine because they were in a space beyond the new airshaft; the airshaft

* Information regarding the nineteen survivors was taken from Forbes, et al., *Report*, appendix F. However, some spellings have been altered by the author, as gravestone readings and other verification methods contradicted spellings in the *Report*. The document stated that sixteen were taken to Sheltering Arms Hospital. However, if one went to Cherrington and three were returned to their homes, that would make for a total of twenty. The discrepancy was not accounted for.

diverted much of the poisonous gas away from the area where the men were because it provided the most direct route for the gases to escape.[39] Newspaper accounts credit John Dean, mine foreman, who was among the men behind the barricade, for saving the lives of his men. In his report to J.J. Forbes, Walter H. Tomlinson stated that Dean and others had made several attempts to get through the afterdamp to the new airshaft but failed and then decided to erect a barricade; this decision probably saved their lives.[40] However, despite what the survivors said and the newspapers published in praise of Dean, those who did the investigation afterward expressed some hesitance to congratulate Dean. Daniel Harrington, a chief engineer at the Safety Division of the Bureau of Mines, stated:

> *It occurs to me that we should be careful to secure as definite information as may be feasible as to what actual service Mr. Dean, and possibly others did towards the removal of men from underground at or after the explosion, in order that data may be available for the J.A. Holmes award committee in the near future.*
>
> *The clippings have in them a considerable amount of interesting information, and also a considerable amount of pure "bunk."*[41]

Handwritten on this letter was the following added by J.J. Forbes:

> *How about proposing Marshall* [sic] *not all the details up* *** [indecipherable] *will shoot for Marshall. I do not think Dean is in line.*[42]

Near 10:00 p.m. and after the nineteen men were taken under the care of doctors, the rescue team made its way back to the location where the bodies of W.E. Tytus and his employees and guests had been found. They were joined by Bureau of Mines representatives J.J. Forbes, supervising engineer; G.W. Grove, mining engineer; and nine others. Forbes and Grove had driven to Millfield from Philadelphia and arrived around the same time as the rescue train.[43] The other members of the bureau either arrived with them, drove in their own vehicles or came on car no. 3 along with the equipment and supplies. The nine men were as follows: M.J. Ankeny, H.R. Burdelsky, S.P. Howell, K.L. Marshall, J.O. Marshall Jr., Raymond A. Morgan, Walter H. Tomlinson, W.D. "Dan" Walker Jr., and J.M. Webb. The bureau team entered the mine and met with rescue teams. After a brief consultation, the expanded team of rescuers and bureau personnel proceeded northward on roadways 5 and 6 and turned

Above: The Ohio National Guard prepares to assist with survivors and remove the dead (view 1). *Photographer and source unknown.*

Left: The Ohio National Guard prepares to assist with survivors and remove the dead (view 2). *Photographer and source unknown.*

onto 21 and 22 west. At the face of 21 west (see illustration "Findings of the Rescuers"), another barricade was found. Behind it, the bodies of seven men were discovered, apparently having been dead for some time.[44] At around the same time, an additional Ohio deputy mine inspector, Thomas Reese, found his way to the rescue party.[45] The team then explored what remained unchecked of 5 and 6 north and 21 and 22 east. When they completed that task, they went to 17 west at the intersection of 4 north, where they were met by a rescue team from Elm Grove Mine, Valley Camp Coal Company, Elm Grove, West Virginia…all wearing gas masks.[46] The Elm Grove team explored 4 north and went onto 19 west.[47] There, they found seventeen-year-old Andy Kish Jr., "the trapper boy, at his assigned station at the sand box at the west end of the switch on 19 west."[48] A crew referred to as a "fresh air crew" came along behind the Elm Grove group and discovered the body of Clyde Dean. Both Dean and Andy Kish Jr. were reportedly badly burned.[49]

As the clock moved toward midnight, rescuers believed they had found all of the deceased miners. For the remainder of the night, crews comprising volunteer miners, national guardsmen, students and community members were tasked with bringing up the bodies and taking them to the makeshift morgues on-site.*

Morticians faced an overwhelming amount of work and were assisted by volunteer nurses. In an equipped mortuary, the average time it takes to prepare a body for burial when there is no external damage is approximately 2 hours. Given that many of the miners did have such damage, it is unknown how long the morticians worked. Even at the minimum of 2 hours per body (not counting the bodies that had to be identified by loved ones or have death certificates completed before processing), there were at least 156 hours of work to be accomplished so that families could bury their dead in a timely manner. It was difficult enough to have seventy-eight bodies to work with, but by midnight, it was clear that there were four more to be found.

> *Identification of bodies began late Wednesday in the first morgue set up in a Millfield billiard room. On Thursday about noon the SCCC Coal Co.*

* County Coroner Lewis F. Jones arrived on scene, rented two rooms (a pool hall and a store in Millfield) for ten dollars each and used the company store owned by the Sunday Creek Coal Company as temporary morgues "to provide morticians with places to wash the bodies, to examine them, and to carry on the necessary steps to identify them." Jones would work many hours over the three days and nights it took to complete the work. In addition to his work at the mine, he made many trips to the towns of Athens, Nelsonville, Trimble and Glouster to get the huge number of supplies he and the undertakers needed in order to do their jobs.

Office and storeroom and an empty storeroom known as "Stack's" (located on the main street in Millfield) were turned into improvised headquarters for the 20 some undertakers who were called into service.[50]

ACTIONS ABOVEGROUND THAT FIRST DAY

As mentioned previously, many activities were occurring aboveground simultaneously with the rescue efforts that were ongoing in the mine. Organizations were called in for support. Many individuals quickly recognized the immensity of the catastrophe and wanted to volunteer to do whatever they could to alleviate the suffering of the families. Help poured in from locals and local agencies, as well as from state and national agencies.

In Nelsonville, principal [Lewis E.] *Buell went to the classrooms to ask if there were any Boy Scouts or former scouts. A number were later transported to the scene at Millfield. Their purpose was to assist the nurses in such as* [sic] *setting up cots and keeping hot water bottles hot.*[51]

Dr. Joseph. M. Higgins, Athens County Health Commissioner, went to Millfield shortly after the explosion. Nurses from the Red Cross, the Health Office in Logan and the Sheltering Arms Hospital in Athens made immediate arrangements for any miners who might be brought out of the mine alive. Cots from ambulances were placed in two temporary hospitals.

Food for the miners and relief workers was sent from Athens to Millfield and distributed by Red Cross volunteers. An Athens restaurant reports that it furnished 215 gallons of coffee and more than 1,000 sandwiches Wednesday night. A call for more food was sent into the Red Cross office at 9 o'clock that night. Women prepared sandwiches with bulk fillings such as cheese and ham and made coffee at the Home Economics Department of Ohio University. The company store at Millfield is furnishing food for the families.[52]

Volunteers from The Salvation Army and The American Legion assisted in the preparation and delivery of food and other support tasks.

Twenty-five nurses from Columbus hospitals were rushed to Millfield Wednesday night in charge of Dr. Drew Davis of Columbus. These nurses assisted the morticians in caring for the bodies after they were brot [sic] *out of the mines and taken to the temporary morgues in Millfield.*[53]

Among the greatest needs of those who were attempting to rescue miners were appropriate masks and breathing apparatus, which were, apparently, in short supply or not easily accessible. This limited the ability of early rescuers to get into the mine safely and retrieve the men they hoped might still be alive. A rescue car (train hauled) had been requested of the Bureau of Mines to bring in supplies and equipment; however, it was stationed in California, Pennsylvania, and would take some time to make its journey. In a letter to J.J. Forbes sometime after the incident, Raymond A. Morgan described the efforts made to move a rescue car as quickly as possible to the disaster site. The station agent at California, Pennsylvania, was notified at 2:35 p.m. that a "special move" was requested for Rescue Car #3. The car left California at 3:24 p.m. All equipment was tested while the train was en route and readied for use on arrival.[54] Morgan explained that the train arrived at Pittsburgh at 4:32 p.m., and while there, a change of crew and engine occurred. It continued on the main line of the Pennsylvania Railroad to Trinway, Ohio, where, once again, a crew and engine change was required. The train then continued on the branch line as far as New Lexington, Ohio, where it was transferred to the New York Central Railroad at 8:23 p.m. However, it didn't arrive at Millfield until 9:40 p.m. In other words, the train made the haul from California, Pennsylvania, to New Lexington, Ohio (approximately 150 miles), in what was considered record time in 1930: five hours, with three stops along the way. However, the shortest distance remaining in the trip—between New Lexington and Millfield (approximately twenty-eight miles)—which should have taken less than half an hour, took over an hour.

> *While en route over the Pennsylvania Railroad, the car was stopped but three times, twice to change crews and engines and once at Denison, Ohio, to refill water tanks on the car. However, while enroute on the New York Central, the car was stopped four times in twenty-eight miles, once it was switched for a period of about ten minutes for a passenger train to pass.*[55]

Unfortunately, despite the efforts of the Bureau of Mines and the Pennsylvania Railroad, by the time the train arrived, almost all of the bodies had been found, and nineteen survivors were in the process of being resuscitated. The personnel sent along with the train car were helpful in the investigation that took place afterward, but the needed equipment was too late to make a difference in the rescue operation.

THURSDAY, NOVEMBER 6, 1930

By 5:00 a.m. on Thursday, all but four men had been accounted for.[56] By 6:00 a.m., the last of the bodies that had been found the previous day were aboveground and were in the process of being prepared for burial or were awaiting the work of the morticians and their assistants. Forbes and Grove had returned to the surface, where plans for further recovery operations were being developed.[57]

Walter Tomlinson states in his report that on that day, he, M.J. Ankeny and H.R. Burdelsky, along with two Ohio deputy mine inspectors and company officials, entered the mine at about 5:00 a.m. (probably a typographical error, as others reported 7:00 a.m.) to search for the four missing men and to get into a portion of the mine that had not been explored the previous day.[58] Ankeny, in his two-page report of the same venture, states that he, Walter Tomlinson, Dan Walker, H.R. Burdelsky and four laborers went into the mine at 7:00 a.m., and their purpose was to search for "the body of the pumper" that had not been recovered.[59] (Dan Walker's report agrees with Ankeny regarding the 7:00 a.m. launch of the search team.)[60]

Ankeny provides his recollections of the trek that day and the lingering danger of poisonous gases. For example, at one point, team members detected a significant amount of carbon monoxide in the portion of the tunnel they were in. The men, not having gas masks, were directed to an area where they were certain there was fresh air. When the group, which had previously been divided, came together again at 17 west near 3 and 4 north, they made decisions about moving the air to ventilate the northern section of the mine.[61] At one point, the men "discovered that a heavy content of carbon monoxide was in the atmosphere" where they were working. As the men moved quickly away from the space, one of the canaries succumbed to the gas. Once they were safe, the team communicated with State Inspector Smith, who told them the body of the pumper had been identified among those that had been brought up the previous evening. With that, the entire team returned to the surface.[62] The plan—as Tomlinson presented it—had included getting into a section of the mine not yet explored. However, after an all-day shift, the team had not been able to find any of the four missing men.[63] Walker suggests that the work ended at 3:00 p.m.[64]

That afternoon or evening (reports differ), Supervisor J.J. Forbes called together state officials (unidentified), mine inspectors, the people representing the Bureau of Mines and company officials for the purpose of planning the remaining work to be done—specifically to get the remainder of the mine ventilated and find the four remaining victims.[65] The attendees settled on an agreement that three shifts of workers would be created consisting of two Bureau of Mines representatives, two Ohio state inspectors and representatives of the Sunday Creek Coal Company. Additionally, six to twelve material handlers, brattice men and drivers would be on each of the three shifts for eight hours each, "working continuously beginning at 7:00 o'clock, Friday morning and continuing until the work was completed."[66] The work would involve erecting brattices and doors and analyzing the new airshaft so the still-unexplored regions of the mine—both active and inactive—could be accessed. A major goal of the teams would be to get to the new airshaft while protecting themselves from the ongoing danger of poisonous gases that could harm them.

ACTIONS ABOVEGROUND ON THURSDAY

Aboveground, families planned the funerals of their loved ones. Bodies were being claimed and removed. Emotional strain was likely on the faces of everyone involved. Lack of sleep, trauma, confusion and disbelief were undoubtedly wreaking havoc. Grief was widespread. One can only imagine the mental state of the widows who lost not only husbands but sons as well—all in a flash of time. No one knows what Mary Williams was going through as she gave birth that day—the day after her husband died—to her fifth child. Not only did she have a newborn who would, out of necessity, require her full attention, but she also had to think about the burial of her husband and the very real issue of caring for four other small children without a means of income.

Volunteer work continued. Fred Spaulding, representing Ohio's Red Cross, had arrived on Wednesday. However, as he observed the work underway through that evening and into Thursday morning, he was so confident of the work being done by the Athens chapter under the leadership of F.D. Forsyth, the group's secretary, that he thought there was little he could add. Before he left that morning, he stated, "The work is progressing in such an orderly manner that I do not think it is necessary for me to stay here."[67]

An article in the *Glouster Press* asserted that the needs of families were assessed.[68] Clothing was distributed. Food was supplied for families through the company store, with the express command of the Sunday Creek Coal Company that "all should be looked after." The article goes on to say that the Columbus Lions Club arranged for flowers to be distributed to homes as tributes to the dead miners and in support of their families who might be without flowers when the dead were returned to them. A floral spray was sent to each home.

John L. Lewis, on behalf of the United Mine Workers of America, sent a check for $2,000 in care of David Fowler of Columbus, with the agreement that the monetary gift should be used for relief work.[69] The gift was quite magnanimous considering that union workers had lost their battle in the previous year's strike and were now working in a nonunion shop and contributing no union dues.

Organizations such as the Red Cross were fully aware of the social calamity that faced the widows and children of the dead miners, and they were doing what they could to anticipate how they might be helpful in the future to counterbalance the families' loss of income.

Friday, November 7, 1930

Tomlinson's team again entered the mine and proceeded toward the new airshaft, where they found a body, which turned out to be that of Phil Powell, one of the fire bosses.[70] In his report, Tomlinson states that Powell had apparently cut a hole in board stoppings that separated a mine opening from the new airshaft and tried to crawl through but was overcome by gas and died in his final effort to escape. After finding Powell's body, the team continued toward the new substation and found a second body in the return airway side of the tunnel. This man, like Powell, appeared to have been trying to reach the surface through the new airshaft. Eventually, the last two bodies were discovered on a stairway inside the shaft.[71] The bodies were removed by midnight.[72]

While others worked to remove bodies, the search team continued in its attempts to force carbon monoxide from the mine and examine the areas still unchecked. The perplexing ventilation problem would not be resolved until the following day.

Saturday, November 8, 1930

Saturday was spent finalizing the plan for regulating proper airflow. The 11:00 p.m. to 7:00 a.m. crew completed the repairs. Ventilation was properly regulated, and preparations were made for the formal investigation that was to occur after their shift ended Sunday morning. The crew that completed the ventilation work included Dan Walker, who reported that he went into the mine with two others from the Bureau of Mines: a Mr. Andrews and a Mr. Brown, who represented the State of Ohio, and eighteen Sunday Creek Coal Company employees.[73]

On this day, families were able to withdraw $112.50 for each person killed; this money had been set aside previously for the specific purpose of providing families with a way to pay for burials.[74] The sum of $9,225 had been previously deposited in a local bank to be used for this purpose.

SUNDAY, NOVEMBER 9, 1930

The official underground investigation was launched at 10:55 a.m. on Sunday, November 9, and was completed at 4:20 p.m. Taking part in the investigation were representatives of the mining company, the State Mining Department and the U.S. Bureau of Mines. Results were included in the final report.[75] Apparently, the goal was to try to determine, based on evidence and their various degrees of expertise, what happened to cause the explosion and to determine the safety of the mine for workers to return to their responsibilities for extracting coal. Though all had surmised that the arcing of a live electric wire on a steel rail had ignited methane and then coal dust, there was a need to determine whether the mine was safe once ventilation had been restored and what needed to be done to prevent such an event in the future. The group reported that the principal damage to the mine was caused by the blowout of stoppings; dropped trolley and power lines in certain sections; the displacement of a pump near 19 west and 4 north, located in room 4 off 18 west; and the displacement of I beams on 19 west off 5 north. Some track was also apparently displaced on 21 and 22 east off 3 north. The total aggregate of damage to the mine itself was ultimately calculated to be $1,500.00 (the equivalent of $27,248.17 in 2023). Added to that figure is $9,225 ($167,576.27 in 2023) for burial expenses and widow funds (for which no final figure could be found). However, the most anyone could collect for a death was $6,500.00 ($118,434.67 in 2023); thus, a maximum—if all received full funding (which they didn't)—would have been $533,000.00 ($9,682,184.37 in 2023). Of course, there was also a fifteen-day loss of production. How much of this money was paid by insurance and how much was paid directly by the company is unknown. Still, one cannot help but wonder how much money it would have taken to remove the likelihood of the disaster happening in the first place.

NOVEMBER 9–20, 1930

November 9, 11 and 13 were spent taking dust and air samples to assess the role coal dust played in the explosion and to consider recommendations for reopening the mine and keeping it safe for returning workers. On November 12, the coroner's inquest was held at the movie theater in Millfield and was conducted by Coroner Lewis F. Jones, who was assisted by Prosecutor Roy D. Williams. On the afternoon of November 14, train car no. 3 left the mine for Pittsburgh.[76] Sunday Creek Coal Company was now under the leadership of George K. Smith, the company's chairman of the board. From November 14 onward, Smith and his staff planned for resuming mine operations as quickly as possible. The mine was reopened on November 20, 1930, fifteen days after the worst mining disaster in Ohio's history.[77]

4

THOSE WHO DIED

Various people have attempted to put together accurate lists of those who died on that fateful day. Several lists have been published with inaccuracies, incorrect names and duplications of names or have included more or less than eighty-two names with no explanation for the discrepancies. A simple verification of death certificates provides the basic information of who the eighty-two men were, but even those documents are flawed for many reasons.

It is quite easy to see why and how mistakes have been made. The inability to read and write was common in many areas of the country in 1930. Some miners had to leave the spellings of their names to those who wrote for them—census takers, legal entities and friends—who were apt to write down phonetic sounds the miners gave for their names, thus leaving trails of various spellings and inaccuracies in the reporting of information. Many miners went by nicknames that, in some cases, had nothing to do with their given names. It was not uncommon for foreign-born workers to attach typical American names to themselves and avoid using their given names; perhaps they did this to "fit in" with their coworkers or just to avoid the mispronunciations and misspellings by their fellow workers or members of the community. As a result, some of those who died have been inaccurately named and have gone unrecognized for who they were. In other cases, the public has been given inaccurate information that has added to the confusion about who actually died in the disaster.

For example, the Millfield Memorial Stone (in Millfield) lists three McGees: C.F., Cam and Earl. There is no "C.F. McGee" who perished on November 5. The memorial lists two people with the last name Sycks; however, only one can claim that last name. The other is actually Charles Szekeres. It is well known that eighty-two men died, yet the stone lists eighty-three. The spellings of names are in some cases inconsistent with the spellings that can be found in census records, the death certificates and the reading of gravestones. Anthony Bycofski went by the names "Andrew Bycofski" and "Andy Cuba"; as a result, he sometimes gets listed as two different people in various lists of the dead. On the Millfield Memorial, he is listed as "Andy Cuba Bycofski." The point here is not to demean the people who erected the monument but to point out how difficult it has been to get accurate information about the men who died.

There has been more than enough confusion about the victims of the explosion. The Keish family (three dead) and the Kish family (two dead) have been erroneously listed as one family in various renditions of the disaster story in newspaper articles and in a book that has been relied on by many people for information since its publication in 1937: *Keeping the Home Fires Burning*, by Damon D. Watkins. His list is woefully inaccurate—to say the least—and his description of the event itself demonstrates that he had more of a flair for fiction than for gathering facts. Newspaper articles provided numerous outlandish stories that demonstrated that the reporters were taking mere hearsay as news without checking for facts. Appendix F of the *Report* lists eighty-one names, rather than eighty-two, and provides names of men who were not killed in the mine and omits the names of some who were. It also lists five Kish family members. It is easy to see why there has been difficulty coming up with the correct names and information. Only through a careful review of death certificates was it possible to begin to acquire an accurate list of those who died. However, names didn't always lead to accuracy about the individual men. Death certificates were filled out quickly on-site and, in some instances, provided little more than a name and what someone identifying the body might volunteer to tell the person who was completing the form.

For example, according to the person who gave the information for John Green's death certificate, Green was born in "Russia" (which could, in reality, mean he could have been born in a Slavic country outside Russia); his birth name is not known. The person who identified the body told the coroner that Green's father's name was John Wassell. Whether or not that was accurate is anyone's guess. Attempts to find a "John Green" or a John

Wassell who in any way matched this miner or his father in the census records or elsewhere were unsuccessful. Finding information on Urban Horvath (often listed as "Hovath") was also unsuccessful. According to local historian William E. Peters (1857–1952), George Keish was actually "Jurgis Jusskiewick." How Peters came by this information is unknown—as is how he came by that particular spelling. Genealogical research did not find such a spelling. In keeping with typical translations of names, the more likely spelling would have been "Juszkiewicz" (which at least appeared as a recognized surname for people from the various ethnic groups within Russia). However, that didn't help in tracking the family or this particular miner. The miner known as Walter Andryvich has been listed in numerous ways. He and his brother had both been in the military, where their names were recorded as "Andrychewicz"; yet locally, the name was interpreted as "Andryvich" (which ended up on Walter's tombstone), "Anderwich," "Andervich," "Undervitch," et cetera. Miners who came from societies that used Cyrillic script (Russian or Slavic), even if they could write their names in Cyrillic, would likely have had difficulty translating their names to English spellings; thus, others did it for them—sometimes very badly.

The following list started with a careful look at the death certificates for each of the eighty-two miners who died. Then, extensive genealogical research was done to verify the men's dates of birth, birthplaces, parentage, marriages, offspring who were living in the home at the time of each miner's death, military service and burial locations, as well as correct spellings of names where such information was available.* Where possible, several resources were consulted to overcome conflicting information. It is the author's belief that this list will clarify many questions about the eighty-two men who died and serve as a starting point for those who wish to do further genealogical research on them.

ANDREWS, ROY BURTRUM. Miner. Born on February 25, 1882, in Buchtel, Ohio, to Paul Andrews and Mary (Watson) Andrews. He married Ethel Phillips on September 12, 1908, in Athens County, Ohio. At the time of his death, there were four children in the home: Lucy, fourteen; Paul, twelve; Rolland, ten; Mary, one. Age at death: forty-eight years, eight months and ten days. He was buried at Green Lawn Cemetery in Nelsonville, Ohio, on November 10, 1930. Death Certificate no. 63874.

* Certainly, many of the miners had grown children who no longer lived in their households. Recorded here are only those who were actually living in the home at the time of the disaster.

ANDRYVICH/ANDRYCHEWICZ, WALTER. Miner. Born on June 25, 1896, in Malanowo, Gmina Brochów, Sochaczew, Poland, to Wojciech Andrychewicz and Josephine (maiden name unknown). He married Alma Jones in 1921 at Parkersburg, West Virginia. At the time of his death, there were three children in the home: Elsie Marie, seven; Kenneth R., three; Wanda J., one month. Military service: army; enlisted on May 26, 1918; became sergeant in 1919; discharged in May 1919. Age at death: thirty-four years, four months and eleven days. He was buried at Hill Top in Millfield, Ohio, on November 9, 1930. Death Certificate no. 63847.

BAUER, DELMAR L. Miner. Born on February 15, 1911, in Ohio to John Bauer and Cora (Crabtree) Bauer. He was single. Age at death: nineteen years, eight months and twenty days. He was buried at Horeb Cemetery in Oak Hill, Ohio, on November 8, 1930. Death Certificate no. 63852.

BAUER, JOHN M. Miner. Born on August 25, 1871, in Ohio to Chris Bauer and Melinda (Metzler) Bauer. He was widowed (married Cora Belle Crabtree on March 17, 1891, in Jackson County, Ohio). At the time of his death, there were three children in the home: Delmar*; Estaline, sixteen; Juanita, fourteen. Age at death: fifty-nine years, two months and ten days. Buried at Horeb Cemetery in Oak Hill, Ohio, on November 8, 1930. Death Certificate no. 63851.

BERGIN, JOSEPH A. Superintendent, Ohio Power Company. Born on July 29, 1886, in Detroit, Michigan, to Frank Bergin and Mary (Rigney) Bergin. He married Estelle M. Hogan on July 16, 1912, in Detroit, Michigan. At the time of his death, there was one child in the home: Robert F., seventeen. Age at death: forty-four years, three months and seven days. He was buried at Brighton Cemetery in Brighton, Michigan, on November 10, 1930. Death Certificate no. 63886.

BROWN, GEORGE WILLIAM. Miner. Born on March 26, 1899, in Perry County, Ohio, to W.P. Brown and Emma (Colburn) Brown. He married Nina Williams Stillwell on May 12, 1928, in Athens, Ohio. At the time of his death, there were no children in the home. Age at death: thirty-one years, seven months and nine days. He was buried at Green Lawn Cemetery in Nelsonville, Ohio, on November 10, 1930. Death Certificate no. 63867.

BROWN, SAMUEL COLBURN. Miner. Born on September 17, 1901, in Perry County, Ohio, to W.P. Brown and Emma (Colburn) Brown. He married Esther Leffler on October 8, 1921, in Athens, Ohio. At the time of his death, there were no children in the home. Age at death: twenty-nine years, two months and eighteen days. He was buried at Green Lawn Cemetery in Nelsonville, Ohio, on November 10, 1930. Death Certificate no. 63859.

BROWN, WILLIAM ROBSON. Miner. Born on October 5, 1881, in Hocking County, Ohio, to Robert Brown and Jane (Robson) Brown. He married Katharine Zigler on September 27, 1905, in Athens County, Ohio. At the time of his death, there were no children in the home. Age at death: forty-nine years, one month and zero days. He was buried at Maplewood Cemetery in Glouster, Ohio, on November 10, 1930. Death Certificate no. 63894.

BURDISS, PAUL JAMES. Miner. Born on December 16, 1892, in Jacksonville, Ohio, to George Burdiss and Kathrin (Gross) Burdiss. He married Lena Exenkamper on May 15, 1920, in Glouster, Ohio. At the time of his death, there were two children in the home: George, seven; Ernest, four. Military Service: navy; enlisted on April 10, 1917; discharged on May 29, 1919. Age at death: thirty-seven years, ten months and twenty days. He was buried at West Union Street Cemetery in Athens, Ohio, on November 10, 1930. Death Certificate no. 63888.

BURNICK, ALEX. Miner. He was born in 1878 in Hungary. He was married (his wife's first name and maiden name are unknown). Age at death: fifty-five years. He was buried at Green Lawn Cemetery in Nelsonville, Ohio, on November 12, 1930. Death Certificate no. 63883.

BUTSKO, JOHN. Miner. He was born on April 24, 1903, in Congo, Ohio, to Mr. and Mrs. Andrew Butsko. He married Dorothy Mayles on May 10, 1930, in Perry County, Ohio. At the time of his death, there was one unborn child in the home: John Thomas (born on January 12, 1931). Age at death: twenty-seven years, six months and eleven days. He was buried at New Lexington Cemetery in New Lexington, Ohio, on November 10, 1930. Death Certificate no. 63870.

BUTSKO, JOSEPH. Miner. He was born in March 1908, in Congo, Ohio, to John Butsko and Magdalan (Chordish) Butsko. He was single. Age at death:

twenty-two years. He was buried at New Lexington Cemetery in New Lexington, Ohio, on November 8, 1930. Death Certificate no. 63846.

BYCOFSKI, ANTHONY/ANDREW "ANDY CUBA." Miner. He was born on July 13, 1887, in Poland to Mr. and Mrs. Jacob Bycofski. He married Ella Roback on January 19, 1909, in Buchtel, Ohio. At the time of his death, there were five children in the home: Stephen, twenty; Florence, eighteen; Martha, eight; Andrew Jr., six; Robert, four. Age at death: forty-three years, three months and twenty-two days. He was buried at Queen of Heaven Cemetery in Trimble, Ohio, on November 10, 1930. Death Certificate no. 63845.

CLANCY, MIKE. Miner. Born on October 26, 1865, in Ohio to Mr. and Mrs. Patrick Clancy. He married Abbie Keplar on June 12, 1888, in Hocking County, Ohio. At the time of his death, there were no children in the home. Age at death: sixty-five years and nine days. He was buried at Green Lawn Cemetery in Nelsonville, Ohio, on November 11, 1930. Death Certificate no. 63868.

CLANCY, WILLIAM "BILLY." Miner. Born on March 23, 1894, in Athens County, Ohio, to Mike Clancy and Abbie (Keplar) Clancy. He married Anna Knight on April 30, 1918, in Athens, Ohio. At the time of his death, there were two children in the home: Annabelle, nine; William, six. Age at death: thirty-six years, seven months and twelve days. He was buried at Green Lawn Cemetery in Nelsonville, Ohio, on November 11, 1930. Death Certificate no. 63865.

COEN, PEARL A. Vice-president, Sunday Creek Coal Company. Born on September 26, 1877, in Old Washington, Ohio, to James E. Coen and Katharine (James) Coen. He married Clara O. Heiner on October 4, 1900, in Guernsey County, Ohio. At the time of his death, there were no children in the home. Age at death: fifty-three years, one month and ten days. He was buried at Green Lawn Cemetery in Nelsonville, Ohio (specific date unknown). Death Certificate no. 63843.

DAVIS, FRANCIS R. "FRANK." Miner. Born in August 1886 in Ohio to John J. Davis and Jane (Ellis) Davis. He was single. Age at death: forty-two years and three months. He was buried at West Union Street Cemetery in Athens, Ohio (specific date unknown). Death Certificate no. 63848.

DEAN, CLYDE KASSON. Miner. Born on March 10, 1891, in Lodi Township, Athens County, Ohio, to Kasson Dean and Katherine (Riley) Dean. He married Lily M. Starkey on December 26, 1911. At the time of his death, there was one child in the home: Bergene, seventeen. Military service: army; enlisted on July 22, 1918; private in Company E, 334th Infantry; fought in the Meuse-Argonne Offensive; discharged on May 17, 1919. Age at death: thirty-nine years, seven months and twenty-five days. He was buried at West Union Street Cemetery in Athens, Ohio, on November 8, 1930. Death Certificate no. 63898.

ERWIN, PAUL. Miner. Born on February 20, 1898, in Vinton County, Ohio, to Frances Erwin and Mattie (McLaughlin) Erwin. He married Myrtle Bail (date and place unknown). At the time of his death, there were seven children in the home: Carl, eight; Donald, seven; Mary, six; Helen, five; Leo, four; Dorothy, two; Paul, ten months. Age at death: thirty-two years, eight months and fifteen days. He was buried at Wilkesville Cemetery in Wilkesville, Ohio, on November 10, 1930. Death Certificate no. 63913.

ERWIN, PHIL. Miner. Born on May 5, 1908, in Vinton County, Ohio, to Frances Erwin and Mattie (McLaughlin) Erwin. He was single. Age at death: twenty-two years and six months. He was buried at Wilkesville Cemetery in Wilkesville, Ohio, on November 10, 1930. Death Certificate no. 63912.

ERWIN, SILAS. Miner. Born on August 1, 1910, in Vinton County, Ohio, to Frances Erwin and Mattie (McLaughlin) Erwin. He was single. Age at death: twenty years, three months and four days. He was buried at Wilkesville Cemetery in Wilkesville, Ohio, on November 10, 1930. Death Certificate no. 63911.

FIELDER, BENJAMIN HARRISON. Fire boss. Born on October 23, 1891, in Lawrence County, Ohio, to George Fielder and Elizabeth (Rose) Fielder. He married Goldie Conrad on July 12, 1910, in Athens, Ohio. At the time of his death, there were three children in the home: Lauretta, eighteen; Florence, sixteen; Willard, thirteen. Age at death: thirty-nine years and twelve days. He was buried at Green Lawn Cemetery in Nelsonville, Ohio, on November 9, 1930. Death Certificate no. 63854.

GREEN, JOHN. Miner. Born in 1884 in Russia to Mr. and Mrs. John Wassell (as reported by the informant for the death certificate). Age at death: forty-six years. He was buried in Hill Top Cemetery in Millfield, Ohio, on November 9, 1930. Death Certificate no. 63921.

GRIMM, CHARLES. Miner. Born on March 26, 1912, in Millfield, Ohio, to Miles Grimm and Mary (Parker) Grimm. He was single. Age at death: eighteen years, seven months and nine days. He was buried at Maplewood Cemetery in Glouster, Ohio, on November 10, 1930. Death Certificate no. 63890.

GRIMM, DAVID MILES. Miner. Born on September 7, 1877, in Sutton Township, Ohio, to David Grimm and Margaret (King) Grimm. He married Mary Jessie Parker on September 9, 1900, in Athens County, Ohio. At the time of his death, there were three children in the home: Charles*; William, fourteen; Minnie, eleven. Age at death: fifty-three years, one month and twenty-nine days. He was buried at Maplewood Cemetery in Glouster, Ohio, on November 10, 1930. Death Certificate no. 63889.

HARLEY, THOMAS ANDERSON. Mine boss. Born on November 20, 1889, in Pennsylvania to John Harley and Mary (Anderson) Harley. He married Effie Bickel on July 25, 1912, in Marietta, Ohio. At the time of his death, there were no children in the home. Age at death: forty-one years, eleven months and sixteen days. He was buried at Ohler Cemetery in Rutland, Ohio, on November 9, 1930. Death Certificate no. 63844.

HAYDEN, WALTER JOHN. Superintendent of Mine no. 6. Born on March 23, 1879, in Minorsville, Meigs County, Ohio, to John Hayden and Mary (Grimm) Hayden. He married Julia A. Schnur on February 11, 1905, at Glouster, Ohio. At the time of his death, there were no children in the home. Age at death: fifty-one years, seven months and thirteen days. He was buried at Maplewood Cemetery in Glouster, Ohio, on November 9, 1930. Death Certificate no. 63895.

HILLEN, JOHN HENRY. Miner. Born on June 3, 1907, in Athens County, Ohio, to Thomas Hillen and Isabell (Johnson) Hillen. He married Mildred Mayles on January 11, 1930. At the time of his death, there were no children in the home. Age at death: twenty-three years, five months and two days. He was buried at Hill Top Cemetery in Millfield, Ohio, on November 9, 1930. Death Certificate no. 63869.

HOOPS, CHARLES. Miner. Born on October 19, 1903, in Shawnee, Ohio, to Jackson Hoops and Permelia (Jones) Hoops. He married Doris Dunkle on December 24, 1926, in New Lexington, Ohio. At the time of his death, there was one child in the home: Jennice, two. Age at death: twenty-seven years and sixteen days. He was buried at New Lexington Cemetery in New Lexington, Ohio, on November 8, 1930. Death Certificate no. 63853.

HORVATH, URBAN. Miner. Born in 1886 in Russia. Age at death: forty-four years. He was buried at Hill Top Cemetery in Millfield, Ohio, on November 9, 1930. Death Certificate no. 63915.

HUNTER, CHARLES. Miner. Born on January 26, 1900, in Lathrop, Ohio, to Pearl Hunter and Mary (Lenigar) Hunter. He was single. Age at death: thirty years, nine months and nine days. He was buried at West State Street Cemetery in Athens, Ohio (specific date unknown). Death Certificate no. 63871.

HUNTER, RAY EDWARD. Miner. Born on December 13, 1906, in Nelsonville, Ohio, to Pearl Hunter and Mary (Lenigar) Hunter. He was single. Age at death: twenty-four years, ten months and twelve days. He was buried in West State Street Cemetery in Athens, Ohio, on November 10, 1930. Death Certificate no. 63877.

HURD, JAMES. Miner (loader). Born on November 22, 1904, in Sand Run, Ohio, to James Findley Hurd and Anna (Bernon) Hurd. He married Lillian McManaway on November 28, 1923, in Athens, Ohio. At the time of his death, there were two children in the home: Adrian, four; Maynard, two. Age at death: twenty-five years, eleven months and thirteen days. He was buried at Green Lawn Cemetery in Nelsonville, Ohio, on November 9, 1930. Death Certificate no. 63902.

JACKSON, GEORGE JOSEPH. Miner. Born on November 6, 1909, in Canaanville, Ohio, to George Jackson and Mary (McCormick) Jackson. He was single. Age at death: twenty years, eleven months and twenty-nine days. He was buried at West Union Street Cemetery in Athens, Ohio (specific date not known). Death Certificate no. 63917.

JENNICE, JAMES. Miner. Born on January 28, 1894, in Italy to Mr. and Mrs. Santo Jennice. He married Mary Domenico on June 12, 1919, in Glouster,

Ohio. At the time of his death, there were six children in the home: Margaret, ten; Irene, nine; Albert, seven; Lewis, six; Charles, four; Virginia, three. Age at death: thirty-six years, nine months and twenty-three days. He was buried at Queen of Heaven Cemetery in Glouster, Ohio, on November 9, 1930. Death Certificate no. 63906.

KEISH, GEORGE (JUSZKIEWICZ, JURGIS). Miner. Born in 1878 in Russia. He married Anna (maiden name and date and place of marriage are unknown). At the time of his death, there were ten children in the home: William*; Stanley*; Walter, fifteen; Edna, fourteen; Alexander, eleven; Harley, ten; George, nine; Frank, six; Robert, two; Mary A., one. Age at death: fifty-two years. He was buried at Queen of Heaven Cemetery in Trimble, Ohio, on November 9, 1930. Death Certificate no. 63919.

KEISH, STANLEY (JUSZKIEWICZ, STANLEY). Miner. Born in 1913 in Russia to Mr. and Mrs. Jurgis Juszkiewicz (George Keish). He was single. Age at death: seventeen years. He was buried at Queen of Heaven Cemetery in Trimble, Ohio, on November 9, 1930. Death Certificate no. 63918.

KEISH, WILLIAM (JUSZKIEWICZ, WILLIAM). Miner. Born in 1908 in Russia to Mr. and Mrs. Jurgis Juszkiewicz (George Keish). He was single. Age at death: twenty-two years. He was buried at Queen of Heaven Cemetery in Trimble, Ohio, on November 9, 1930. Death Certificate no. 63849.

KERN, FRANK. Miner. Born on December 8, 1878, in Mount Vernon, Ohio, to Mr. and Mrs. John Kern. He married Laura Allen on October 31, 1928, in Vinton County, Ohio. At the time of his death, there was one child in the home: Isabelle R., one. Age at death: fifty-one years, ten months and twenty-eight days. He was buried at West Union Street Cemetery in Athens, Ohio, on November 7, 1930. Death Certificate no. 63849.

KISH, ANDREW, JR. Miner. Born on September 18, 1913, in Bellaire, Ohio, to Andrew Kish Sr. and Esther (Vaske) Kish. He was single. Age at death: seventeen years, one month and seventeen days. He was buried on November 10, 1930, at Hill Top Cemetery in Millfield, Ohio. Death Certificate no. 63866.

KISH, ANDREW, SR. Miner. Born on April 1, 1888, in Hungary to John Kish and Bertha (Block) Kish. He married Esther Vaske (the date and place of marriage are unknown). At the time of his death, there were nine children in the home: Andrew Jr.*; John, fifteen; Elizabeth, thirteen; Gaza, twelve; Alexander, ten; William, nine; Peter, seven; Julia, five; Helen three. Age at death: forty-two years, seven months and four days. He was buried at Hill Top Cemetery in Millfield, Ohio, on November 10, 1930. Death Certificate no. 63880-A.

LANCASTER, HUBERT E. Civil engineer. Born on April 28, 1890, in Nelsonville, Ohio, to William Lancaster and Hannah (Pickle) Lancaster. He married Belle Adamson on June 9, 1915, in Nelsonville, Ohio. At the time of his death, there were two children in the home: Calvin, thirteen; Robert, eleven. Age at death: forty years, seven months and seven days. He was buried at Green Lawn Cemetery in Nelsonville, Ohio, on November 8, 1930. Death Certificate no. 63876.

LEFEVER, WEBSTER EMERSON. Miner. Born on November 1, 1871, in Trimble, Ohio, to Mr. and Mrs. Samuel Lefever. He was a widower (he married Lydia Bell Armstrong on November 12, 1891, in Athens County, Ohio). At time of his death, there were no children living in the home. Age at death: fifty-nine years and four days. He was reportedly buried in Senecaville, Ohio (the burial location and date of burial are unknown). Death Certificate no. 63899.

LOVE, GEORGE. Miner. Born on December 4, 1899, in Athens County, Ohio, to William Love and Nancy (Matheny) Love. He married Dora Ramsey on June 4, 1921. At the time of his death, there was one child in the home: Herman, ten. Age at death: thirty years, eleven months and one day. He was buried at Beach Grove Cemetery in Monroe Township, Perry County, Ohio, on November 8, 1930. Death Certificate no. 63880.

LYONS, JAMES ARCHIE "ARCH." Miner (loader). Born on August 3, 1905, in Goose Run, Ohio, to Charles Lyons and Roxabel (Wade) Lyons. He married Margaret Showalter on August 30, 1929, in Chauncey, Ohio. At the time of his death, there were no children in the home. Age at death: twenty-five years, three months and two days. He was buried at Buchtel Cemetery in Buchtel, Ohio, on November 9, 1930. Death Certificate no. 63903.

MARTIN, JAMES CHESTER. Miner. Born on October 6, 1908, in Ohio to Chester and Elsie Martin. He was single. Age at death: twenty-two years and one month. He was buried at Horeb Cemetery in Oak Hill, Ohio, on November 8, 1930. Death Certificate no. 63850.

MCALLISTER, JOHN "JACK." Miner. Born on August 4, 1887, in Jacksonville, Ohio, to Peter McAllister and Elizabeth (Kane) McAllister. He married Lena (Schall) Chesser Perry on April 21, 1930, in Washington County, Ohio. At the time of his death, there were two stepchildren in the home: Claude Chesser; eighteen; Kenneth Chesser, fifteen. Military service: army; enlisted on July 22, 1918; private; Company K, 138th Infantry; he served in the Meuse-Argonne Offensive; discharged on May 17, 1919. Age at death: forty-three years, three months and one day. He was buried at Maplewood Cemetery in Glouster, Ohio, on November 8, 1930. Death Certificate no. 63905.

MCGEE, EARL JAMES. Miner. Born on September 22, 1906, in Glouster, Ohio, to Cam McGee and Louise (Davis) McGee. He married Ethel Anderson on February 12, 1926, in Athens, Ohio. At the time of his death, there were two children in the home: Earl Jr., three; Billie, twenty-two months. Age at death: twenty-four years, one month and fourteen days. He was buried at Maplewood Cemetery in Glouster, Ohio, on November 10, 1930. Death Certificate no. 63891.

MCGEE, JAMES CAM. Miner. Born on November 2, 1879, in Pennsylvania to William Harvey McGee and Anna (Yeats) McGee. He married Louise Davis on July 2, 1905, in Glouster, Ohio. At the time of his death, there were two children in the home: Edith, twenty-two; George, twenty. Age at death: fifty-one years and three days. He was buried at Maplewood Cemetery in Glouster, Ohio, on November 10, 1930. Death Certificate no. 63892.

MCKEE, ELLSWORTH FLOYD. Miner. Born on August 26, 1900, in Athens County, Ohio, to Wesley McKee and Anna (Rutter) McKee. He married Lucille Baker on March 5, 1930, in Chauncey, Ohio. At the time of his death, there was one child in the home: Leslie Elroy, three months. Age at death: thirty years, two months and ten days. He was buried at Hill Top Cemetery in Millfield, Ohio, on November 9, 1930. Death Certificate no. 63860.

MCLEAN, GEORGE ANDREW BARRIS. Miner. Born on November 19, 1889, in Coshocton, Ohio, to Adam and Sarah McLean. He married Lenna Hoisington on December 13, 1913, in Amesville, Ohio. At the time of his death, there were three children in the home: Adam, fifteen; Sarah, twelve; Wilber, four. Age at death: forty years, eleven months and five days. He was buried at Augustine Cemetery in Sugar Creek, Dover Township, Ohio, on November 8, 1930. Death Certificate no. 63907.

MCMANAWAY, EDWARD, JR. Miner. Born on February 13, 1903, in Orbiston, Ohio, to Edward McManaway Sr. and Lessie (Robison) McManaway. He married Goldie Milliron on December 19, 1921, in Logan, Ohio. At the time of his death, there was one child in the home: Robert H., five. Age at death: twenty-eight years, eight months and twelve days. He was buried at Green Lawn Cemetery in Nelsonville, Ohio, on November 11, 1930. Death Certificate no. 63879.

MESSENGER, JOHN WELLINGTON "WILLIAM." Miner. Born on June 4, 1885, in Gilmer County, West Virginia, to John M. Messenger and Elizabeth (Radcliff) Messenger. He married Susie Coleman on March 12, 1921, in Athens, Ohio. At the time of his death, there were seven children in the home: Clifford, seventeen; Emma E., fourteen; George, twelve; Goldie M., eight; John, six; Madge, five; Louisa J., one. Age at death: forty-five years, five months and one day. He was buried at Hill Top Cemetery in Millfield, Ohio, on November 9, 1930. Death Certificate no. 63884.

MILLIRON, HARRY. Miner. Born on March 24, 1902, in Burr Oak, Athens County, Ohio, to Ephraim Milliron and Almeda (Hatfield) Milliron. He married Essie Petitt on September 13, 1923. At the time of his death, there were two children in the home: Loraine, five; Evelyn, three. Age at death: twenty-eight years, eight months and nineteen days. He was buried at Green Lawn Cemetery in Nelsonville, Ohio, on November 8, 1930. Death Certificate no. 63900.

NADROSKI, JOHN. Miner. Born on February 27, 1907, in Wheeling, West Virginia, to Joseph Nadroski and Frances (Knicik) Nadroski. He was single. Age at death: twenty-three years, eight months and twenty-four days. He was buried at Mount Calvary Cemetery in Athens, Ohio, on November 10, 1930. Death Certificate no. 63910.

NORTH, JAMES E. Miner. Born on December 13, 1890, in Zanesville, Ohio, to Henry North and Eva (Perry) North. He married Zora Parker on August 28, 1910, in Jacksonville, Ohio. At the time of his death, there were five children in the home: Wilbur*; June, fifteen; Vern, twelve; Ruth, ten; James four. Age at death: thirty-nine years, ten months and twenty-three days. He was buried at Glouster Cemetery in Glouster, Ohio, on November 8, 1930. Death Certificate no. 63872.

NORTH, WILBUR. Miner. Born on June 27, 1911, in Athens County, Ohio, to James North and Zora (Parker) North. He was single. Age at death: nineteen years, four months and eight days. He was buried at Glouster Cemetery in Glouster, Ohio, on November 8, 1930. Death Certificate no. 63873.

PARRY, WILLIAM RAYMOND. Miner (loader). Born on June 24, 1893, in Nelsonville, Ohio, to William Parry and Jane (Grose) Parry. He married Lena Ferrel on August 25, 1917, in Ohio. At the time of his death, there were two children in the home: Geraldene, six; Marilyn, four. Age at death: thirty-seven years, four months and twelve days. He was buried at Green Lawn Cemetery in Nelsonville, Ohio, on November 9, 1930. Death Certificate no. 63904.

PARSONS, ROBERT DAY. Vice-president and manager of the Columbia Chemical Division of the Pittsburg Plate Glass Company. Born on August 21, 1885, in Akron, Ohio, to William Cheny Parsons and Sarah Day (Seymour) Parsons. He married Dorothy Galt on October 23, 1915, in Akron, Ohio. At the time of his death, there were two children in the home: Hugh Galt, thirteen; Robert Day Jr., eleven. Age at death: forty-five years, two months and fifteen days. He was buried at Rose Hill Cemetery in Akron, Ohio, on November 8, 1930. Death Certificate no. 63897.

PATTERSON, JOHN W. Miner. Born on October 15, 1905, in Ward Township, Hocking County, Ohio, to Roy Patterson and Emma (Christian) Patterson. He married Evelyn Spearry on December 11, 1926, in Ohio. At the time of his death, there was one child in the home: Everett "Pat," three. Age at death: twenty-five years and twenty-one days. He was buried at Hill Top Cemetery in Millfield, Ohio, on November 10, 1930. Death Certificate no. 63858.

PETITT, FLOYD DAVID. Miner. Born on July 16, 1890, in Greenup County, Kentucky, to John Petitt and Susan (Holly) Petitt. He married Gladys Lanning on September 13, 1915, in Hocking County, Ohio. At the time of his death, there was one child in the home: Delbert, thirteen. Age at death: forty years, three months and nineteen days. He was buried at Green Lawn Cemetery in Nelsonville, Ohio, on November 8, 1930. Death Certificate no. 63857.

PEYATT, THOMAS WESLEY. Miner. Born on September 29, 1908, in Coal Township, Jackson County, Ohio, to Grant Peyatt and Margaret (Lohr) Peyatt. He was single. Age at death: twenty-two years, one month and seven days. He was buried at Hill Top Cemetery, Millfield, Ohio, on November 9, 1930. Death Certificate no. 63882.

PEYATT, WILLIAM HENRY. Miner. Born on October 9, 1905, in Coal Township, Jackson County, Ohio, to Grant Peyatt and Margaret (Lohr) Peyatt. He was single. Age at death: twenty-five years and twenty-seven days. He was buried at Hill Top Cemetery in Millfield, Ohio, on November 9, 1930. Death Certificate no. 63881.

PHILLIPS, VIRGIL. Miner. Born on December 11, 1907, in Floodwood, Ohio, to John Oliver Phillips and Blanch (Hankison) Phillips. He was single. Age at death: twenty-three years, ten months and twenty-four days. He was buried at Bethel Holiness Mission Cemetery in Knox, Vinton County, Ohio, on November 8, 1930. Death Certificate no. 63856.

POWELL, PHILLIP. Miner and fire boss. Born on August 21, 1883, in Nelsonville, Ohio, to Phillip Powell and Elizabeth (Maffin) Powell. He married Alice Skiver on December 15, 1906, in Nelsonville, Ohio. At the time of his death, there were six children in the home: Francis, eighteen; Donald, thirteen; Willow, eleven; Phillip, ten; James, eight; Richard, four. Age at death: forty-seven years, two months and fifteen days. He was buried at Green Lawn Cemetery in Nelsonville, Ohio, on November 11, 1930. Death Certificate no. 63885.

RANKIN, ABRAHAM. Miner. Born on October 26, 1874, in Lawrence County, Ohio, to William Rankin and Eliza Delilah (Sharp) Rankin. He married Emma Ashworth on July 15, 1897, in Lawrence County, Ohio. At the time of his death, there were no children in the home. Age at death: fifty-six years and ten days. He was buried at Hill Top Cemetery in Millfield, Ohio, on November 9, 1930. Death Certificate no. 63875.

ROBERTS, VERNON LYLE. Foreman, Columbia Cement Plant. Born on November 10, 1893, in Newton Township, Muskingum County, Ohio, to

Charles Lincoln Roberts and Daisy (Lyle) Roberts. He married Anna Grace Phillips on June 9, 1917, in Fultonham, Ohio. At the time of his death, there were two children in the home: Lyle, twelve; Betty, ten. Age at death: thirty-six years, eleven months and twenty-five days. He was buried at Fultonham Cemetery in Fultonham, Ohio, on November 9, 1930. Death Certificate no. 63887.

ROBINETT, CARL D. Miner. Born on November 13, 1900, in Jacksonville, Ohio, to Francis "Frank" Marion Robinett and Nora (Judson) Robinett. He married Helen Lucille Gillespie on October 20, 1923, in Chauncey, Ohio. At the time of his death, there were two children in the home: Stewart V., five; Victor, two. Age at death: twenty-nine years, eleven months and twenty-two days. He was buried at Nye Cemetery in Chauncey, Ohio, on November 9, 1930. Death Certificate no. 63855.

SYCKS, HARRY JACOB "JAKE." Miner and wire hanger. Born on September 2, 1889, in Logan, Ohio. He married Maude Palmer on May 31, 1930, in Middleport, Ohio. At the time of his death, there were no children in the home. Age at death: forty-one years, two months and three days. He was buried at Augustine Cemetery in Sugar Creek, Dover Township, Ohio, on November 10, 1930. Death Certificate no. 63914.

SZEKERES, CHARLES. Miner. Born in 1870 in Cseke, Hungary, to Mr. and Mrs. Albert Szekeres. He married Mary (her maiden name and their date and place of marriage are unknown). At the time of his death, there were no children in the home. Age at death: sixty years. He was buried at Hill Top Cemetery in Millfield, Ohio, on November 8, 1930. Death Certificate no. 63916.

THOMAS, GEORGE. Miner. Born on January 22, 1907, in New Pittsburg, Hocking County, Ohio, to Perry Rice Thomas and Donna (Fisher) Thomas. He married Shirley Willette Ruth on July 3, 1929, in Chauncey, Ohio. At the time of his death, there were no children in the home. Age at death: twenty-three years, nine months and thirteen days. He was buried at Green Lawn Cemetery in Nelsonville, Ohio, on November 8, 1930. Death Certificate no. 63864.

TONAK, ANDY. Miner. Born in 1862 in Hungary. He was married (his wife's name is unknown). At the time of his death, there were no children in the home. Age at death: sixty-eight years. He was buried at Hill Top Cemetery in Millfield, Ohio, on November 10, 1930. Death Certificate no. 63878.

Trainer, Thomas Brush. Traffic manager, Columbia Cement Company. Born on May 25, 1891, in Zanesville, Ohio, to Thomas S. Trainer and Adelaide (Fogle) Trainer. He married Marie F. Blandy on September 4, 1915, in Zanesville, Ohio. At the time of his death, there were no children in the home. Age at death: thirty-nine years, five months and ten days. He was buried at Woodlawn Cemetery in Zanesville, Ohio, on November 8, 1930. Death Certificate no. 63896.

Tytus, William Ewing. President, Sunday Creek Coal Company. Born on April 17, 1890, in Middletown, Ohio, to John B. Tytus and Minnesota (Ewing) Tytus. He married Frances Jones on September 25, 1920, in Granville, Ohio. At the time of his death, there were three children in the home: John S., eight; Alice, five; William Jr., two. Military Service: first lieutenant, air service, from the enlisted reserve corps; enlisted on December 12, 1917; made captain on October 3, 1918; First Aero Squadron (until October 15, 1918); Allied expeditionary force in France until 1919; honorable discharge on August 28, 1919. Age at death: forty years, six months and nineteen days. He was buried at Woodside Cemetery in Middletown, Ohio, on November 8, 1930. Death Certificate no. 63842.

Upson, Howard Henry. Division manager, Sunday Creek Coal Company. Born on May 18, 1888, in Newark, Ohio, to Howard Upson and Celia (Palmer) Upson. He married Juliet D. Besuden on April 17, 1917, in Newark, Ohio. At the time of his death, there were three children in the home: Juliet, ten; Howard, nine; Annette, seven. Age at death: forty-two years, five months and eighteen days. He was buried at Cedar Hill Cemetery in Newark, Licking County, Ohio (the specific date of his burial is unknown). Death Certificate no. 63841.

Wade, Alfred Nelson. Miner. Born on January 1, 1861, in Lawrence County, Ohio, to David Wade and Nancy (Large) Wade. He married Lottie Tolliver on July 6, 1916, in Athens, Ohio. At the time of his death, there were no children in the home. Age at death: sixty-nine years, ten months and four days. He was buried at Canaanville Cemetery in Canaanville, Ohio, on November 9, 1930. Death Certificate no. 63909.

WADE, LUTHER. Miner. Born on January 31, 1900, in Milton Township, Jackson County, Ohio, to Alfred Nelson Wade and Ellen (Wilbern) Wade. He married Alice Wright in 1919 in Parkersburg, West Virginia. At the time of his death, there were six children in the home: Alfred, nine; Garnet, eight; Glenna, six; David, five; Luella, three; Marion, one. Age at death: twenty-nine years, ten months and five days. He was buried at Canaanville Cemetery in Canaanville, Ohio, on November 9, 1930. Death Certificate no. 63908.

WEISS, JOHN. Miner. Born on April 2, 1884, in Buchtel, Ohio, to John Weiss and Wilhelmina (Horenburg) Weiss. He married Florence Kittle on November 28, 1906, in Athens County, Ohio. At the time of his death, there were two children in the home: Mildred E., eighteen; Ruben Edward, fifteen. Age at death: forty-six years, seven months and three days. He was buried at Maplewood Cemetery in Glouster, Ohio, on November 11, 1930. Death Certificate no. 63893.

WILLIAMS, JOHN WESLEY. Miner. Born on May 15, 1900, in Hollister, Athens County, Ohio, to Samuel Williams and Della (West) Williams. He married Mary Elizabeth McConaha on August 13, 1921, in Trimble Township, Athens County, Ohio. At the time of his death, there were five children in the home: Violet, seven; Claude, six; Betty L., four; Garnet, two; John, born on November 8, 1930. Age at death: thirty years, six months and seven days. He was buried at Glouster Cemetery in Glouster, Ohio, on November 9, 1930. Death Certificate no. 63901.

WILLIS, ANDREW. Miner. Born on December 25, 1904, in Hamden, Vinton County, Ohio, to Oscar Willis and Minnie May (Bolton) Willis. He was single. Age at death: twenty-five years, ten months and ten days. He was buried at Green Lawn Cemetery in Nelsonville, Ohio, on November 9, 1930. Death Certificate no. 63863.

WILLIS, OSCAR. Miner. Born on August 11, 1884, in Milton Township, Jackson County, Ohio, to Andrew Willis and Mary (Sherman) Willis. He married Minnie May Bolton on August 13, 1904, in Wellston, Ohio. At the time of his death, there were six children in the home: Andrew*; Virgil*; Charles, twenty-three; Hazel, thirteen; Gerald, nine; Kathleen, six. Age at death: forty-six years, two months and twenty-five days. He was buried at Green Lawn Cemetery in Nelsonville, Ohio, on November 9, 1930. Death Certificate no. 63861.

WILLIS, VIRGIL. Miner. Born on April 13, 1911, in Hamden, Vinton County, Ohio, to Oscar Willis and Minnie May (Bolton) Willis. He was single. Age at death: nineteen years, six months and twenty-three days. He was buried at Green Lawn Cemetery in Nelsonville, Ohio, on November 9, 1930. Death Certificate no. 63862.

COMPENSATION FOR WIDOWS AND CHILDREN

Many of the widows not only had to deal with their personal grief but also had to worry about the costs of burying their dead and, in many cases, caring for young children or aged parents who were still in the home. The families had been told there would be some compensation, but what that compensation would be and how soon it would be available was uncertain in the earliest days after the mine disaster. Hungry children couldn't wait for bureaucracies to reach conclusions. Fortunately, as stated earlier, some relief for funeral expenses was available almost immediately. Compensation for the widows was not handled in the same way. According to one *Athens Messenger* article, representatives from the Ohio Industrial Commission took responsibility for receiving and filing claims; they were stationed at the office of the mine at the company store.[78] The article stated that as a result of the Workmen's Compensation Act, the maximum amount anyone could receive was $6,500, a sum that certainly wouldn't carry a young widow with small children for more than a few years—and that was assuming the cost of living wouldn't increase. In 1930, the average income was $1,368, or approximately $26 per week. It's fairly obvious that widows faced significant financial issues no matter how frugal they may have been. In truth, many did not receive the full amount they were owed, and no one received funds instantaneously.

Two newspaper articles regarding widow compensation provide information about who was paid and how much they were paid. Though neither of the articles account for all the widows involved, they provide some indications of how the money was disbursed.

One of the articles stated that some widows were paid on a weekly basis until their total allotment had been used up.[79] The State Industrial Commission was responsible for processing cases and making payments. What criteria they used for determining the release of funds is unclear. From the outset of reviewing cases, a problem presented itself—families in which not only the father had died in the mine, but the unmarried sons who had still been living in their parents' home had died as well. In particular, this article mentions the George Keish, Oscar Willis and Andrew Kish families. The widows would be allotted money for the loss of their husbands, but a determination had to be made about payments for sons, which, it was decided, would depend on how much the family relied on the income of those sons—undoubtedly, a difficult matter to prove. The following figures dealing with twenty-two individuals come directly from the article; figures could not be corroborated:

Hoops, Doris, widow of Charles Hoops, weekly payments of $18.75 for a total of $6,500.00.
Hurd, Lillian, widow of James Hurd, weekly payments of $13.43 for a total of $5,539.88.
Keish, Anna, widow of George Keish, weekly payments (amount unstated) for a total of $6,500.00.
Kern, Laura, widow of Frank Kerns, weekly payments of $18.75 for a total of $6,500.00.
Love, Dora, widow of George Love, weekly payments of $13.36 for a total of $5,682.56.
Lyons, Margaret, widow of James "Archie" Lyons, weekly payments of $14.41 for a total of $5,994.56.
McLean, Lenna, widow of George McLean, weekly payments of $14.72 for a total of $6,123.52.
Messenger, Susie, widow of William Messenger, weekly payments of $15.62 for a total of $6,500.00.
Parry, Lena, widow of William "Raymond" Parry, weekly payments of $18.75 for a total of $6,500.00.
Patterson, Evelyn, widow of John Patterson, weekly payments of $18.75 for a total of $6,500.00.
Petitt, Gladys, widow of Floyd Petitt, weekly payments of $12.14 for a total of $5,050.24.
Powell, Alice, widow of Phillip Powell, weekly payments of $18.75 for a total of $6,500.00.
Robinett, Lucille, widow of Carl Robinett, weekly payments of $14.86 for a total of $6,181.76.
Willis, Minnie, widow of Oscar Willis, weekly payments of $13.81 for a total of $5,774.00.

The article lists others with totals only, and it is unknown whether these families were paid outright or received a weekly payment. They are as follows:

Andryvich, Alma, widow of Walter Andryvich, $6,500.00.
Brown, Katherine, widow of William Brown, $6,500.00.
Bycofski, Ella, widow of Anthony "Andy Cuba" Bycofski, $6,500.00.
Clancy, Abbie, widow of Michael Clancy, $6,500.00.
Kish, Esther, widow of Andrew Kish, $6,500.00.
Sycks, Maude, widow of Harry Sycks, $6,500.00.
Thomas, Perry, father of George Thomas, $5,553.60.
Wade, Lottie, widow of Alfred Wade, $4,123.28.

Another article provides information about widows (and others) who were paid in lump sums for the loss of their loved ones.[80] The reporter stated that there were some cases that would not be decided on right away. The widows of Wilbur North and Charles Grimm would have to wait for a determination about "the dependence of the mother on the sons," though they could be compensated for the loss of their husbands. Charles Szekeres's widow would have to wait, as the case was referred to the foreign vice counsel to determine "dependency of widow and children in Hungary." The twenty-one recipients listed and the amounts granted them were as follows:

Andrews, Ethel, widow of Roy Andrews, $6,500.00.
Brown, Kate, widow of William Brown, $6,500.00.
Burdiss, Lena, widow of Paul Burdiss, $6,500.00.
Butsko, Dorothy, widow of John Butsko, $6,500.00.
Clancy, Anna, widow of William Clancy, $6,500.00.
Dean, Lillian, widow of Clyde Dean, $6,500.00.
Fielder, Goldie, widow of Benjamin Fielder, $6,500.00.
Grimm, Mary, widow of David Grimm, $5,528.64.
Harley, Effie, widow of Thomas Harley, shared with mother-in-law, $6,500.00.
Jennice, Mary, widow of James Jennice, $5,819.84.
Lancaster, Belle, widow of Hubert Lancaster, $6,500.00.
McAllister, Lena, widow of John McAllister, $6,240.00.
McGee, Ethel, widow of Earl McGee, $5,665.22.
McGee, Louise, widow of Cam McGee, $5,079.36.

McKee, Lucille, widow of Ellsworth McKee, shared with mother and sister who were dependent on Ellsworth, $4,821.44.
McManaway, Goldie, widow of Edward McManaway, $6,500.00.
Milliron, Essie, widow of Harry Milliron, $6,285.76.
North, Zora, widow of James North, $5,408.00.
Rankin, Emma, widow of Abraham Rankin, $6,500.00.
Wade, [Alice] Bessie, widow of Luther Wade, $4,692.48.
Weiss, Florence, widow of John Weiss, $6,500.00.
Williams, Mary, widow of John Williams, $6,227.52.

5
INVESTIGATION

The *Report of Gas and Dust Explosion in Mine No. 6, Sunday Creek Coal Company, Millfield, Ohio, November 5, 1930* is the official record of the disaster. It was based on the reports of Bureau of Mines professionals who aided in the investigation, the findings of the coroner's inquest, the conclusions of the Ohio state mine inspector and scientific evidence related to ignition of the explosion, levels of gases, dust accumulation and airflow. The following people are known to have submitted reports that were used in consideration of the final document: Frank A. Ray, consulting mining engineer, Columbus, Ohio;* Dan Walker, foreman miner, rescue car no. 3, Bureau of Mines;† C.W. Owings, member of the Safety Division, Bureau of Mines;‡ E.W. Smith, chief of Division of Mines (Ohio), Cambridge, Ohio;§ M.J. Ankeny, Bureau of Mines;¶ Raymond A. Morgan, first-aid miner, Rescue Car no. 3, California, Pennsylvania;** and Walter H. Tomlinson, foreman miner, Bureau of Mines, car no. 1.†† Scientific analyses were done by H.M. Cooper, chemist, who assessed the dust samples taken in the mine; William P. Yant, supervising chemist, Bureau of Mines, who assessed gas sampling done at the mine; and D.J. Demorest, a professor of metallurgy at Ohio State University, who

* Undated.
† February 28, 1931.
‡ February 19, 1931.
§ Undated.
¶ Undated.
** December 26, 1930.
†† November 20, 1930.

determined the cause of the spark that detonated accumulated methane. George S. Rice, chairman of the Mine Safety Board, provided editorial feedback to the first draft of the report.

The introduction to the final report claims that the following eleven people from the Bureau of Mines assisted in either "rescue operations or in the investigation": M.J. Ankeny, H.R. Burdelsky, J.J. Forbes, G.W. Grove, S.P. Howell, J.O. Marshall Jr., K.L. Marshall, R.A. Morgan, W.H. Tomlinson, W.D. Walker Jr. and J.M. Webb.

E.W. Smith, in his report, states that the following people were involved in exploring the mine to determine the explosion's cause and starting point (formal investigation):[81]

E.W. Smith, Ohio Division of Mines, Cambridge, Ohio
Andrew Ginnan, deputy mine inspector, Jacksonville, Athens County, Ohio
J.J. Forbes, U.S. Bureau of Mines, supervising engineer, Pittsburgh, Pennsylvania
G.W. Grove, U.S. Bureau of Mines
W.J. Ankeny, U.S. Bureau of Mines
S.P. Howell, explosives engineer, U.S. Bureau of Mines
G.W. Glidden, district engineer, E.D. Bullard and Co.
Howard Ong, engineer, Millfield Coal and Mining Co.
F.S. Knox, Industrial Commission of Ohio
Frank A. Ray, consulting engineer
Loren Holmes, Warren Colliers Co.
William Brown, Ohio Colliers Co.
Barney Clay, Hanna Coal Co.
S.W. Ross, engineer, Sunday Creek Coal Co.
E.W. Bonnett, mining engineer, New York Central Railroad Co.
H.E. Nold, professor, mining engineering, Ohio State University

It should be noted that various people who wrote reports did not agree on the names of the people involved as a whole or those who participated in specific incidents that occurred in the rescue or investigative activities. In some cases, two professionals who were supposedly in the same place at the same time had different recollections of what happened and who was present at the time. In some cases, the actual writing of these reports occurred well after the investigations, and writers likely relied heavily on memories or brief notes, some of which may have been inaccurate. Two of the people who wrote reports appear to have used the opportunity to vent their grievances against the people who preceded them in the efforts to deal with the crisis, as well as with those attempting to gather evidence. For example, in his report to J.J. Forbes—who had the responsibility of collecting data from the various professionals who were expected to give reports—Dan Walker, who was a foreman miner on rescue car no. 3, stated:

> *Practically no means of communication existed between the inside and outside. Consequently, much time lapsed before definite action was taken by bureau men. In the meantime, men, women and children were allowed to crowd about the shaft portal which was also the outlet for the return from one section of the mine. This condition was not only dangerous, but deplorable before action was taken to keep the crowds back. When more men were needed for inside work, they were chosen as volunteers from the surrounding crowd, most of them students and sightseers. Many of these men never reached the point of operations and were a hinderance to those trying to work, were endangering their own life* [sic] *and health and a hazard to all in the mine.*[82]

Raymond A. Morgan, a first-aid miner who had arrived with Walker on car no. 3, also complained about the conditions he found when he arrived:

> *It was apparent that there was very little organization. A crowd of four or five hundred people stood around the collar of the shaft, most of them trying to look down, and many of the men smoking. A few of the National Guard men were attempting to keep the crowd back but were unsuccessful in their efforts. Anyone who would volunteer to help in the rescue work during the night were taken underground and it made little difference if it was their first trip.*[83]

Though the reports that Walker and Morgan wrote will always be part of the record of the Millfield Mine Disaster, it is important—to this author, at least—to challenge the notion that the situation was out of control and that, as they seem to imply, there was very little organization until the bureau men took over.* First of all, communication had been going on all day between the "inside" and the "outside." Robert Marshall, Andrew Ginnan and others moved quickly to notify local, state and national officials to get the

* Their reports are still within an official file; the author was able to obtain copies in 2011. However, the tone of those reports was tempered by those involved in compiling information from all the various reports submitted and preparing the final report, *Report of Gas and Dust Explosion in Mine No. 6, Sunday Creek Coal Company, Millfield, Ohio, November 5, 1930.*

professionals at the site as quickly as possible while the two men organized a crew to go in search of miners who might be injured or trapped. When supplies for building brattices were necessary, the rescue crew was supplied with what they needed. Aboveground, support agencies were being called and medical support had been contacted and were either on-site or making their way through the traffic congestion to get there. It could easily be argued that given the magnitude of the disaster, the availability of people with experience in dealing with such a calamity and lack of essential equipment, a number of people with a grasp of what was needed did as much as they could throughout the day while waiting for more experienced professionals to arrive. E.W. Smith, a man capable of effective leadership, arrived in the late afternoon, but he, too, without a team of experienced professionals, was limited in what he could do to oversee all the various activities taking place at the mine site.

Both Walker and Morgan arrived on the scene sometime after 9:40 p.m., as did the other "bureau men" whom Walker alludes to. That Walker criticizes what he found on arrival and the lapse of time seems odd. If it took some time "before definite action" occurred by the bureau men, who is it who is being criticized for not doing what the bureau men did? Who could have or should have taken control and dealt with the myriad issues going on both aboveground and belowground? Apparently, no team of knowledgeable leaders had arrived on the scene with the skills and abilities necessary to handle all the complexities of a disaster site, and yet, for the most part, all that could be done without proper equipment for mine rescue work and for preparation of dealing with the dead and wounded was being done. Walker and Morgan were correct in stating that many people had been allowed to "crowd about the shaft," and they were correct that it was not safe and did not help the situation. However, that it happened is completely understandable. At the time that Walker and Morgan arrived on the scene, people had been standing and waiting all day, hoping for a miracle that would return their loved ones from the mine safe and sound. The national guard had been called in to help, and perhaps the guardsmen could have been more ruthless in clearing people from the landing area, but they were working under the orders they were given.

It is doubtful that any of the national guardsmen or anyone who had assumed any authority wanted to further traumatize grieving people by pushing them like cattle into a holding pen. Yes, there were people who took up space out of curiosity, but then, how were the national guardsmen to identify the motivations of all the people who arrived at the scene? Not only were the few authorities aboveground dealing with endless decisions about getting tasks accomplished, but they were also dealing with a huge number of emotionally charged human beings. Undoubtedly, each family member and friend waiting to hear the fate of their loved ones had difficulties that day with managing their lives and controlling their emotions. The last thing the guardsmen needed was a confrontation with the crowd, which grew throughout the day.

Frances Tytus, the wife of W.E. Tytus, was in Cleveland when she received word of the explosion. She knew that her husband had gone to Millfield to do an inspection and was told that he was in the mine when it happened. Because she had the financial means, she chartered a plane—and somehow found a way to get Mrs. Howard Upson on the plane with her—and flew to Millfield in the midafternoon to stand with the other wives and loved ones who were waiting anxiously to know what had become of the men below.[84] Most of the wives of the miners had already been at the mine long before Mrs. Tytus and Mrs. Upson arrived. Reporters who managed to get to the scene early observed and later painted pictures with their words of the ever-diminishing hopes of those who waited for their loved ones to be brought up on the main shaft's hoist. It is not in any way surprising that the families of the men who died might crowd the landing area, perhaps even testing the ability of the national guard to withstand their collective power. It is not difficult to imagine parents, siblings, wives and children wanting to see each time the hoist was lifted that their loved ones would be coming up alive and well. It is also not difficult to imagine the wailing and unleashed emotional responses as bodies began to be lifted up in the hoisting shaft—perhaps pleading and demanding such things as, "Let me see him! I've got to know if it's him!" Undoubtedly, amid this emotional upheaval, people felt they could identify their loved ones sooner than the coroner would; a coroner likely had no idea what each man looked like. Being told they had to wait wouldn't have done much to keep the crowd in check. In times of extreme stress, people are unlikely to think about much more than their own terror and might not understand why the rescuers and undertakers might try to keep the bodies away from them. The gruesome disfigurations and burns of some of the men, combined with layers of dirt and coal dust in the mine made worse by the violent explosion, made many of the bodies almost unrecognizable—certainly not images anyone would wish for loved ones to see.

Given that nineteen survivors had been found shortly before the bureau men arrived, it is reasonable to think that the families were desperate to know who the survivors were as well. Despite Walker's claim that no communication existed between the crews aboveground and those below, it is likely that there was some communication between the people going in and out of the mine to assist the rescue crews. It is likely that the materials for brattice-making were called for, food and drink for the rescue crews were needed and equipment—when it became available—had to be taken into the mine. It is also likely that someone was required to relay needs and information from the rescuers to people on the surface for any number of reasons. Couriers may have given notice to friends and family members aboveground that survivors had just been found. Certainly, when the onlookers saw doctors being taken into the mine, they would have suspected there was a good chance that survivors had been found. Hope is a powerful motivator, and undoubtedly, many within the crowd hoped desperately that their loved ones would be among the survivors.

Photographs taken on the day of the explosion show that at some point a single rope or cable was strung at waist height to cordon off the landing area from the crowd. Most people honored it; however, it was unlikely to be of much use when people became frantic about their loved ones. Walker states that the crowding occurred "before action was taken to keep the crowds back." How would he know that? Action had been taken: the national guard had been called in and was on scene. Whatever it was that finally constituted taking acceptable action that might have satisfied Walker or Morgan is not explained in their reports, and nothing of the kind was included in the final report.

Both men complained about the volunteers who were taken into the mine to help recover the bodies and claimed that helpers were more trouble than help. However, they offer no examples or explanations to support their claims and give no alternative that would have been more effective. If the two report writers had been able to identify a cadre of people who were capable of being more effective than the people who were on scene when they arrived, they failed to explain who they were.

Despite the fact that many people in charge of the Sunday Creek Coal Company and mine no. 6 were dead, all of the various agencies that needed to know what had happened were contacted quickly. A rescue effort was launched as soon as possible by Andrew Ginnan and Robert Marshall, even though they did not have all the resources they needed at their disposal. Once E.W. Smith arrived, he took charge of the situation until officials from the Bureau of Mines and others could assist. Sunday Creek Coal Company officials were on hand as well, though in what capacity they led the efforts is unknown. Somehow, as has been previously shown, the early rescue crew was getting information about others coming into the mine to join in the rescue efforts. Doctors were called into the mine to assist with caring for the nineteen men who were found alive. A temporary hospital had been set up. Someone was coordinating the delivery of food and coffee to the rescuers, providing the sustenance they needed while they did their work. The Ohio National Guard had been called in to help with crowd control. Local law officers were assisting with traffic control. The call for undertakers had been made, and many responded. Considering the problems associated with travel at that time and the effort it took to sort out all of the logistics of dealing with this disaster, which was unlike anything the local community had ever faced or prepared for, it seems unfair for Walker and Morgan's proclamations to go unchallenged.

THE CORONER'S INQUEST, NOVEMBER 12 IN MILLFIELD

Coroner Lewis F. Jones swore in Pete McKinley, a former deputy sheriff, who had been working at the mine at the time of the explosion. McKinley was hired to serve subpoenas on thirty-six witnesses, some of whom were state mine inspectors who had been called to another explosion in Coshocton, Ohio. Because of the enormity and complexity of the disaster, Jones asked for the assistance of Prosecutor Roy D. Williams. The inquest was conducted in the movie theater in Millfield, the largest indoor space available in the town. All attempts to locate a complete transcript of the coroner's inquest have failed; thus, "restatements of conclusions" included in the *Report* of the Bureau of Mines must be relied on.* The report lists seven "salient features" that were claimed to have come from that proceeding (the punctuation and spelling have been corrected):[85]

A. *That after the pre-shift inspection, the fire bosses worked on the brattices and doors (i.e., attempted to improve the ventilation at the working places).*
B. *That on occasion when the fire bosses told the mine foreman they were going to inspect the old workings, he ordered them to do other things in line with their work: that is, build brattices, etc.*
C. *That the gas accumulation could probably not accumulate in the 3 and 4 north sections between 3:00 a.m. and the following midday.*

---

* The author was able to locate four pages of testimony from Howard Davis from files that had been collected since 2011, when research began. However, the source by which they came into the author's hands is not known.

D. That because of the apparent lack of coke formation, it was thought by some witnesses that the explosion was of gas only, though Deputy State Mine Inspector Andrew Ginnan testified that it was a "gas and dust" explosion.

E. That the explosion originated in the 3rd and 4th north sections inby 20 west off main north.

F. That the most probable cause of the explosion was an electric arc at the point where there was a broken trolley line on 21 east off 3 north.

G. That it was possible that the explosion was caused by an open light.

Both Raymond Morgan and Dan Walker reported their dissatisfaction with the inquest in the way it was conducted and what it provided as evidence. Whether their criticisms were fair could be debated if a complete transcript of the inquest were available; however, what is interesting in their reports of the event are their beliefs that witnesses were, in essence, withholding truth in order to protect the company.

> *My impression of the inquest was that it was useless as far as finding out the true cause of the disaster or if it could have been prevented. One of the reasons for this opinion was that the men who knew anything about the mine could shift the responsibility to one of the officials, who was killed or intimate they did not know. Very few important facts were brought out at the inquest because the man doing the questioning was a lawyer in place of a man familiar with mining. It was evident that everyone who should have known something about the conditions, ventilation, and other factors that might have had a bearing on the cause of the explosion were* [sic] *protecting the Company and was reluctant to tell what they knew of the cause of the disaster. The people who wished to talk were workmen and knew nothing of the ventilation, power distribution or any of the other factors that might have caused an accumulation of gas and its ignition.*[86]

> *The inquest did not disclose any information but left the impression that all concerned were anxious to let the affair drop with just as little publicity as possible. It also brought out the fact that the Chief Inspector or some other competent representative of the State Department should conduct the examination of all persons giving testimony.*[87]

Findings of Investigators

The following items are taken from individual reports of various investigators and the final report, as well as extant letters, memos and telegrams between interested parties.

Force of the Blast and Flame

E.W. Smith, Ohio's chief mine inspector, described the effect of the explosion on the mine.[88] He observed that at a couple of locations, eight-inch I beams had been blown out, one of them bent at a thirty-degree angle; coal cars were blown off rails and wrecked; a five-foot-long section of track rail somewhere off of 19 west was torn from its place and blown ten feet away; a two-inch suction line valve had its handle blown off, and the valve stem was bent by the force; at the face of 22 west, tracks were torn up; all brick stoppings between 3 and 4 north near 22 west were blown out; heavy falls (unspecified as to what had fallen) had occurred at the face of both 3 and 4 north entries. Smith said the "heavy falls" appeared to have taken place after the explosion); and trolley wire had been torn down in various locations.

> *The length of the extreme diameter of the explosion North and South in the 3 and 4 north was about 1,800 feet, and east and west in the 21 and 22 entries, about 1,900 feet. The area affected by the violence of air waves from the explosion was much greater.*[89]

The nearest person to the blast was 1,500 feet away from the ignition point.[90]

Gas and Potential for Explosion

Several reports pointed to four specific problems that likely played a role in the disaster or should have been resolved before sending workers into the mines: (1) Sunday Creek Coal Company Mine no. 6 had long been known to be "gassy" from the time of its creation as Poston Mine no. 6; it had been declared so by the Ohio State Department of mines.[91] (2) Mine workers wore open-flame carbide lamps on their hats*; workers using open lights (flame)

* Open-flame carbide lamps were designed for use in nongaseous mines.

and pellet powder, combined with gas, made for a dangerous situation.[92] (3) Fire bosses carried impermissible key lock flame safety lamps; those lamps were not locked at the time of the accident.[93] (4) Fire bosses were making only cursory pre-shift examinations of the mine and then were occupied mostly in building brattices.[94] Methane accumulation was not detected because, as Owings explains, "the mine foreman had told the fire bosses to spend their time on 'more important' work, such as erecting brattices rather than inspecting old workings."[95] In his report, Owings said the following:

> *The employment of fire bosses in building brattices, doors, etc., (work which can be done by laborers) rather than having them occupied in the much more essential work of inspecting as to safety, is little short of a crime, yet it is more or less customary not only in Ohio but in nearly all other states where fire bosses are employed. The fact that this is a more or less universal practice does not by any means excuse or justify its existence.*[96]

How the Methane Accumulated

> *The 3 and 4 north section* [of the mine] *had been idle for about 6 months prior to November 5, 1930, but the power had been left on the trolley wire. A change had been made in the haulage road about 6 weeks before the explosion, which necessitated the removal of a brattice, and material for installing automatic doors had been brought to this point, but the doors had not been erected. The removal of the brattice and failure to erect the doors caused a short circuit of the air from 3 and 4 north section. It is believed that this condition allowed methane to accumulate gradually, and the accumulation was not detected as the mine foreman had told the fire bosses to spend their time on "more important" work such as erecting brattices rather than inspecting old workings.*[97]

> *Just why such a body of methane was allowed to accumulate in the mine cannot be determined, as both fire bosses were killed in the explosion. The mine foreman makes affidavit to the effect that the 21 and 22 west, 21 and 22 east and face of the third and fourth north entries were all well-ventilated and that he had been over those entries four days before the explosion and there was no standing gas in any of them at that time, and that the fire bosses had been ordered to examine these entries each morning the same as though they were live workings.*

If this statement, which was made under oath, is true, there can only be two conclusions: the first is that the fire bosses failed to carry out their orders, or, if they made the examination as he claims they were ordered to do, there must have been a sudden outflow of gas on the morning [of] *the explosion.... This last seems to be a rather remote theory, and we do not believe it could have occurred. There are many rumors in circulation as to the accumulation of this body of gas, but when we attempt to verify them, they are either denied or found to be only gossip without any tangible proof of their accuracy. While we feel that there was carelessness somewhere, we are unable, with the evidence we have, to say where that responsibility rests.*[98] [Note: alterations in punctuation and grammar have been made for readability and to avoid multiple intrusions of (sic) in these paragraphs.]

About six weeks prior to the explosion, the track and trolley at the junction of 3 and 4 north and 19 west entries had been changed... [making] *necessary the removal of a brattice on 19 west near 3 north.*

Shortly after this, fixtures for an automatic door to be installed at this point had been brought in, but the door had not been made or installed prior to the explosion. This caused a short circuit of air at this point and permitted the accumulation of gas in 3 and 4 north section inby 19 west.[99]

No evidence was found to show that a door had been placed on 19 west off main north between 3 north and 4 north, which was necessary to properly ventilate 3rd and 4th north section. This section, therefore, for several days prior to the explosion, was without ventilation. This section produced from 0.34% to 1% gas which accumulated in this section while it was without ventilation.[100]

Coal Dust and Need for Rock Dusting

Various investigators commented on the need for rock dusting to help manage huge amounts of coal dust in the mine, some of it highly explosive.

Arc-wall coal-cutting machines produced large quantities of fine coal dust, which readily suspended in the air current.[101]

This coal, when in the form of dust, is explosive when suspended in the air... The designation of the Poston No. 6 Mine [Sunday Creek Coal

Company, mine no. 6] *explosion as a "Gas and Dust Explosion," and not simply a gas explosion, is founded on the copious supply of coked particles of dust deposited throughout and beyond the entire explosion area and an estimation of the relative quantities of coked particles, as made by the coke analysis section of the Bureau, in all dust samples taken.... Much dry coal dust was observed throughout the mine, especially in regions affected by the explosion.*[102]

The dust samples within the flame area indicate an ample quantity of fine dust whose total combustible varied from 46.9 percent for read sample A-56264 to 75.7 percent for rib and roof sample A-56267.[103]

It is my further opinion from study of the gas conditions, that coal dust must have been the principal cause of propagation because of the relatively low percentage of methane shown by the gas samples. The absence of coked dust on the timbers and ribs is not evidence that a dust explosion had not occurred, especially where it was started by gas, as seems to be evident in this case.[104]

Impermissable Equipment

Impermissible machinery was used throughout the mine.[105]

Main and secondary hauling was done with trolley locomotives, and gathering haulage was done with crab-reel locomotives.[106]

No permissible machinery of any kind was used underground. The power for all motor-driven machines was 200 to 275 volts DC. There were 9 mine trolley motors, 2 mine line H. motors, 1 switch haul motor, 6 gathering motors of the cable road type, 5 gathering pumps—of which two discharge to the surface, 5 mining machines over cutting, 2 mining machines undercutting, 1 Goodman Universal and 1 Universal.... [Of the seven mining machines] *one of the Jeffrey Arcwall machines at one time was permissible but was not maintained in a permissible condition.*[107]

Large proportion of coal [was] *loaded by Northern Conveyor and Mfg. Company, Janesville, Wisconsin....Each loader has* [a] *1.5 H.P. non-permissible motor.*[108]

Other Concerns

Pellet powder, an impermissible explosive, was used in the mine to blast coal and rock loose. In one section of the mine, experienced people were specifically hired to do shooting (blasting) during their shifts; otherwise, miners elsewhere in the mine were responsible for doing their own shooting at any time they needed to have it done.*[109] *It would seem that having individuals who have little or no training or experience could be quite dangerous to themselves and others. Apparently, boxes of the pellets were left in various places in the mine for easy access.†*

Ventilation was a problem in the mine, which is probably the reason for the company to invest in a new airshaft. Though the motor was capable of circulating about sixty-five thousand cubic feet of air per minute, it apparently was inadequate for the expansions that had taken place since the fan had been installed. Frank Ray, in his report, stated, "It is definitely known that only a small portion of the total intake air reached the active works of this mine."[110]

Improper clearance along haulage ways was noted by Walter Tomlinson.[111]

Condemnations by the Experts

This mine, as appears to be the case with many Ohio mines, was allowed to operate with little regard to safety. Leaving power on wires in abandoned gassy and apparently unventilated workings that were not inspected, using non-permissible explosives and blasting during the working shift, using non-permissible equipment at acknowledged gassy faces, and using open lights in a mine rated by the State as gassy indicates woeful lack of safety or a disregard of well-known dangers.[112]

Conditions inside the mine indicated wanton carelessness on the part of company officials who knew that gas was being produced and that the coal was dry and dusty, no attempt having been made to minimize the danger from this source.[113]

Negligence on the part of the mine management must bear the blame for this explosion. First, because an abandoned section of the mine was allowed to

* Using pellet powder for blasting was referred to as "shooting."

† According to Tomlinson, ten boxes of this powder were found after the explosion, close to where the explosion is believed to have started.

go without ventilation when it was known to generate explosive gas. Second, because live wire was allowed in a portion of the mine abandoned. Third, because no effort had been made to prevent the spread of an explosion by the use of rock dust.[111]

There is no good reason why power should have been on the trolley wire in the inactive 3 and 4 entry section, and still less excuse for power being on the wire prior to the explosion. Lack of attention to these important details made possible the ignition of the body of gas.[115]

From the shaft bottom to the explosion area, conditions indicated bad mining practices and a loosely-run operating organization. Power lines were poorly hung without guard boards for protection of men at crossings. Haulage roads were dirty, without clearance or alignment. Timber, when used, was set back from the track only far enough to clear the cars and motor. Ventilation was on a continuous circuit. The mine is rated as gassy by the Ohio State Department of Mines, fire bosses are employed, and open lights were in use at the time of the disaster.[116]

Judging by the conditions inside the mine after the explosion, it is evident to the writer that the safety of the men was given little consideration by the company officials as evidenced by the following careless practices: a) open lights were used and smoking permitted inside when the mine was known to be gaseous; b) pellet powder was used for blasting; a very dangerous practice where gas is produced; c) miners were permitted to fire their own shots at any and (or) all times during the shift…d) arc-wall coal-cutting machines produced large quantities of fine coal dust which readily suspended in the air current; no attempt was made to minimize the danger from the dust by the application of water or rock dust; e) haulage by trolley locomotive was performed upon return air; wires were hung low and unprotected…wires were placed in return airways liable to contain explosive gas; a substation carrying open live parts was installed underground on return air, and near workings producing gas; f) lack of proper clearance along haulage ways was noticed; also, all roads were found to be exceedingly dirty and dusty.[117]

There are many bad practices at this property, and there is very little regard given to safety. The statement which occurred in the press accounts regarding negligence, of course, is true, but I do not believe it is for the Bureau to engage in publicity of this character.[118]

DISSEMINATING THE RESULTS OF THE INVESTIGATION

Throughout the investigations, the Sunday Creek Coal Company was involved. Representatives from the company were allowed to serve on the teams that engaged in rescue operations and in the investigations themselves. They were told by the Bureau of Mines personnel and representatives from the Ohio State Mine Department what the final report would likely include. George K. Smith, the chairman of the board of the company who had temporarily assumed the role of the deceased president, was told on November 14 that he would have a preliminary report within a few days and a final report in a few weeks and it would be "calling an ace an ace and a spade a spade." The chairman knew the report would shed negative light on the company. However, he also knew that the public could be kept from seeing the results.

Newspaper reporters picked up on and reported some of the failures of the Sunday Creek Company's officials to protect workers. In various articles, they implied there was more to the story than an unfortunate incident that no one could have predicted. However, reporters were largely relegated to telling the stories they could glean from the local people (workers, miners, miners' families and the volunteers who were working to support the rescue teams) or eliciting material from an official who was willing to risk violating the trust of their cohorts and who wouldn't want to be on record until the investigation was (1) completed, (2) reviewed, (3) determined to be accurate and (4) approved for distribution. As a result of little "official" information, the reporters' articles were often inaccurate and contradictory, often causing more confusion than enlightenment.*

The Bureau of Mines team, under the leadership of J.J. Forbes, had the responsibility of compiling all the data that had been acquired, including the reports of various investigators and findings of scientists studying air and gas samples. Analyses and the writing of reports took time. Some reports were not received until well into February 1931. When the *Report* was written, it did nothing to help the general public understand what had happened, how it had happened and who was responsible. It was the policy of the bureau that copies of the final report could be given only to the owners/operators of the mine where the explosion or other disaster occurred. Many people not affiliated with the Sunday Creek Coal Company attempted to receive copies and were consistently refused. In other words, no one could know

---

* This state of confusion has been carried into the present day for many people.

the full story unless and until the company allowed it to be known. One can only imagine what the United Mine Workers leadership might have done with the "official findings" in hand or what attorneys might have done by way of lawsuits on behalf of the people who lost their lives or the loved ones who were left behind. Instead, people who requested any documentation, especially the *Report*, received responses such as the following:

> *I regret that we are not able to comply with your request inasmuch as such reports of mine disasters are held strictly confidential between the Bureau and the mine operator concerned. The Bureau of Mines has no power over the regulations of the operation of privately-owned coal mines; this power being vested in the states and investigation of such disasters by the Bureau can only be made with the consent of the operator. For this reason, the Bureau has adopted the policy of holding reports of such investigations strictly confidential.*[119]

Among the many documents researched for this book, only one mention is made of a formal statement from the Bureau of Mines to the press. A telegram from Daniel Harrington to J.J. Forbes was sent from Washington, D.C., at 12:34 p.m. on November 7, 1930. The message is quoted in its entirety below; however, it has been altered to remove the word *stop* at the end of each sentence and has been converted to upper- and lower-case letters, and punctuation has been added for clarity:*

> *U.S. Bureau of Mines Car No. 3, Millfield, Ohio.*
>
> *Herewith copy of the only statement given by Washington office to press quote by Kenneth Watson, Pittsburgh Press Staff Write Washington* [sic]. *"Daniel Harrington, Chief Safety Engineer for the United States Bureau of Mines announced today* [that] *government field men investigating the Sunday Creek Coal Company's mine disaster reported its miners used open lights despite the fact that Ohio Mine Inspectors had rated it as gassy. 'I have also been informed that the mine failed to use rock dusting,' Harrington said. Rock Dusting is the method prescribed by the government to all mine owners to reduce mine disasters to a minimum through preventing coal dust burning after explosions. The information was obtained by Harrington through a telegram sent him by J.J. Forbes, Superintendent of Operations*

* It was common for many telegrams to be written in all capital letters without use of punctuation. Sometimes the word STOP would be added to indicate the end of a sentence.

for the Bureau's Mine Rescue Station at Pittsburgh and through telephonic conversations Harrington held with officials at the Pittsburgh station. Although use of rock dusting does not prevent explosions entirely, it lessens the dangers of fire and asphyxiation from poisonous gases by reducing the amount of combustion material which might be ignited. The government, although without power to compel use of rock dusting in mines, has for years advocated its use as [a] *safety method. Reports reaching Harrington indicated seventy-seven miners lost their lives in the explosion, that nineteen others were rescued from death by asphyxiation through building barricades which prevented poisonous gas reaching them and that one hundred twenty-seven others had left the mine before the explosion took place." Watson. You will note* [that] *nothing is said as to negligence by anyone. Show this to persons concerned. Sorry that misquotation appeared. Harrington*

Notice that Harrington emphasizes his belief that "nothing is [to be] said as to negligence by anyone." However, the press release ties miners wearing open flame lights to a gassy mine. The fact that the miners wore open flame lights is not portrayed as unsafe or against company rules despite it being widely known that the lights were dangerous.* In fact, the sentence construction would make it appear that the miners chose to use the lights even though they should have known they were dangerous. Is this negligence? No potential alternative for the cause of the explosion is offered other than the miners' lights and presence of dust. Methane is ignored. A lack of rock dusting is mentioned as a failure on the part of the company, but it is rendered a neutral criticism, since the method is only recommended, not required. Whether this was intentional or not, the press release leaves the reader with the impression that the disaster was most likely an unfortunate accident *caused by the miners*, whose open-flame lights inadvertently touched off explosive coal dust, which might have been less problematic had rock dusting been done.

* It is likely that the head lamps were provided by the company. If they were not provided by the company, the head lamps were obviously not prohibited.

6

CONCLUSIONS

Eighty-two men died on November 5, 1930, in an explosion that never should have happened. The vast majority of those men—whose burned, broken or poisoned bodies would be hoisted out of the mine that day and for two days afterward—had worked in an environment with unsafe equipment, inadequate personal protection and unenlightened or deliberately callous leadership. In order to feed their families in very difficult times, miners made do with what was given them and with the circumstances they couldn't overcome, just as their predecessors had done for decades. The proverbial deck was stacked against them, and the one possible means for gaining anything resembling fair pay and safety had once again eluded them when, over the previous two years, they lost the ability to stand together as part of a powerful union and use its resources and organizational skills to demand at least minimal security.

On the day of the explosion, company officials and guests happened to be in the mine with workers and shared their fate: a rarity in mining history. Their loss was a tragedy, too; those men had loved ones who would mourn them just as the miners did. The fact that those company officials and guests had come to inspect two improvements to the operation suggests that they either believed the mine was safe—at least safe enough—or gave it no thought because safety was not high on their lists of concerns. For all concerned, the day was just another day filled with people doing their jobs. However, the mindsets of the mine owners/operators doing their jobs versus mine workers doing their jobs in the mines came from two very different worlds of thought. Mine executives and mine workers have long been at odds—a situation that has caused, in some cases, open warfare between them, needless deaths and political struggles for power and control. Of course, stereotyping is inherently unfair to individuals, but there is a case to be made that mine owners and operators—in general—bought into the notion that mining was about making profits for owners and shareholders and miners should accept what the owners decided was fair in terms of pay and work conditions. The miners generally tended to believe that their back-breaking work was the means for the owners to obtain wealth and that workers deserved a fair share of that wealth as part of their pay—at least enough pay that they could provide adequately for their families. They also sought reasonable safety consideration and some say in the rules that guided their work.

In many ways, the story of coal mining has served as a prime example of one of humanity's greatest conundrums: how to mediate between those who seek or who have gained power to control others and those who wish to be recognized for what they do, treated fairly and respectfully, and valued as human beings rather than chattel. Whether William E. Tytus and/or the people who accompanied him personally saw workers as unfortunate inconveniences is not known.

It should not be surprising to anyone that there has been a long-running battle in the United States' economic and political capitalism-based system over the "rights" of employers and corporations versus the "rights" of workers and citizens: power and oppression versus equality and human worth. Employers, prior to the notion of "minimum wage," believed they should have the right to pay what they were willing to pay and that people who didn't want to accept the pay being offered were free to go elsewhere in search of work that would give them what they wanted to receive, which, of course, was likely to be fruitless, since other potential employers would not be compelled to pay more than other employers. Workers, on the other hand, believed that they should be paid wages that adequately supported themselves and their families proportionate to the risks involved in doing what the company wanted done and the knowledge and skills required to do the work. Underlying the workers' position has always been that workers produce the products and services that enrich their employers and company stockholders and deserve a fair share of the wealth they produce. One side of the issue has the money, connections and power to dominate, while the other side has nothing but its labor and, ultimately, unionization to counter the domination of corporate leadership. Over time, both sides (pro-owner and pro-worker) have had victories and losses, but the underlying battle over the "rights" of the powerful versus the "rights "of the oppressed has continued unabated into the present day and continues to impact the economic and political lives of Americans.

The Millfield mine disaster is a reminder of how the wars between two different views of the world, as described previously, inevitably end. The preponderance of evidence clearly shows that mine no. 6 was known to be dangerous, and its workers were ill-equipped and not properly protected. Unionization had been defeated the previous year, and workers were working for lower wages than they had previously been paid while the mine was under the control of Poston Consolidated Coal Company. The Sunday Creek Coal Company was making a minimal effort to improve some safety issues while it took advantage of increased profit as a result of becoming a nonunionized operation; the company's leaders had used their power to defeat the efforts to pay workers fairly. It was paying what it wanted to pay, managing as it saw fit, dodging anything more than minimal expectations for safety and increasing output while minimizing expenses. In other words, the disaster of November 5, 1930, was just another case of a company's management putting dollars ahead of people, pretending to be sorry when the inevitable happened and suffering little by way of consequences for leaving dead and destroyed people in its wake.

GLOSSARY

afterdamp: A toxic mix of gases that occurs after an explosion caused by firedamp. It is rich in carbon monoxide, carbon dioxide and nitrogen.

airshaft: The vertical opening extending from the land surface to the base of the mine, over which a fan moves fresh air down into the mine.

bituminous coal: There are four main coal classifications—lignite (lowest grade), subbituminous, bituminous and anthracite (highest grade). Bituminous coal is the most prevalent in and around Athens County and most of eastern Ohio. This coal is further graded by its physical and chemical characteristics.

brattice: A partition between columns that aids in directing air flow to desired locations in a coal mine, often made with boards, brick or canvas cloth. Miners make a distinction between the words *brattice* and *stoppings* depending on where the barriers are located and how permanent they are. To avoid undue confusion, the word *brattice* is used herein to denote temporary barriers used for control of air flow.

door: See *mine door*.

entry system: Millfield Mine no. 6 had an entry system referred to as a "double- and triple-entry system." This is a system in which the tract to be exploited is first penetrated by a series of two or more relatively narrow (eight to ten feet) entries (headings or gangways) that are then made relatively wide to access coal through a room and pillar approach to mining.

face: See *mine face*.

fan house: The building that surrounds the motor and fan that is used to provide fresh air to a mine.

firedamp: Any flammable gas in a coal mine. Methane is the most common.

gob pile: Accumulated waste materials removed during coal mining operations.

inby: A direction away from the main entrance (going *into* the mine).

main shaft: The essential access for entering and leaving the underground workings of a shaft mine. See *shaft mine*.

methane: A colorless, odorless and highly flammable gas. It is released from the coal (compressed plant material) and surrounding rock strata.

middle kittanning: A geological identifier for coal seams based on age and distance from coastal marine areas.

mine door: A mine door is used to deflect air from its course in one entry and shift it to another entry, while at the same time permitting the passage of mine cars through the first entry.

mine face: The surface where the mining work is advancing (not yet cut into), like the end of a tunnel that will be extended. Often referred to as "the face."

outby: A direction heading out of the depths of a mine, toward the main shaft/exit.

pellet powder: An explosive similar to black blasting powder, but it is made into pellets and therefore more easily and safely controlled than blasting powder. It is not permissible in gaseous mines.

room and pillar mining: A common method for retrieving coal, particularly in bituminous coal mines. The method involves creating a grid of intersecting passageways through which miners can reach work locations and coal cars can haul extracted coal to the main shaft opening and tipple for processing. Rooms are the spaces made when miners cut coal out of the spaces between the grid lines. Pillars of coal and/or stone are left in place to support the weight of the earth above the underground workings of the mine.

shaft mine: A mine that requires a means for making a vertical entry into a mine. It usually requires an elevator for workers to access the coal deep in the earth. The shaft at mine no. 6 was recorded as being approximately 186 feet deep (various publications provide conflicting information) and required two elevators to handle worker and equipment descents and ascents. It is excavated from ground level downward to the base of the coal seam.

slate fall: Though "slate" is a particular kind of soft stone, the term *slate fall* is used for any kind of stone that falls unexpectedly from the upper part (or the "roof") of a mine tunnel.

substation: An area within the mine where circuit breakers, fuses, switches, transformers, et cetera, are found. A substation is used to control the flow of electrical power from the main power system (aboveground) to the areas being worked in the mine.

switch/switch station: A system of rails whereby coal cars can be redirected onto different tracks.

tipple: The building where coal is "tipped" out of the cars coming up out of the mine for sorting and distribution into railroad cars lined up beneath the tipple.

trolley wire: Overhead electrical lines that power towing machinery (hauling devices) used to move carts filled with coal to the main shaft and to return empty carts to desired locations for filling or for distribution of supplies and equipment.

NOTES

Chapter 1

1. Roy, *History of the Coal Miners*, 12.
2. Ibid., 12.
3. Ibid., 67–8.
4. Ibid., 44–5.
5. Ibid., 230–42.
6. Sharp, "Climax to Come," 1.

Chapter 2

7. "Poston Mine Sets Record," *Athens* (OH) *Messenger*, 12.
8. Ray, "Notes," 3.
9. Ibid.
10. National Register of Historic Places Inventory, "Sunday Creek Coal Company Mine No. 6."
11. Ibid.
12. "Store Is Sold," *Athens* (OH) *Messenger*, 5.

Chapter 3

13. Ray, "Notes," 8.
14. "Men to Probe Depths of the Workings," *Athens* (OH) *Messenger*, 1.
15. Ray, "Notes," 3.
16. Owings, memorandum, 2.
17. Ibid., 1.
18. Ray, "Notes," 6.
19. Owings, memorandum, 1.
20. Smith, "Report of Mine Explosion," 1.
21. Forbes, Grove, Howell and Ankeny, *Report*, 31.
22. Tomlinson, memorandum to Forbes.
23. Ibid., 3.
24. Ray, "Notes," 9.
25. "82 Dead Taken from Millfield Mine 6 Disaster," *Glouster* (OH) *Press*, 1.
26. Forbes, Grove, Howell and Ankeny, *Report*, 31.
27. Forbes, letter to Grove.
28. Forbes, Grove, Howell and Ankeny, *Report*, 32.
29. Ibid.
30. Smith, "Report of Mine Explosion," 1.
31. Ibid., 2.
32. Ibid.
33. Ibid., 3–4.
34. Forbes, Grove, Howell and Ankeny, *Report*, 32.
35. Ibid.
36. Ibid.
37. Ibid.
38. Ibid.
39. Smith, "Report of Mine Explosion," 4.
40. Tomlinson, memorandum to Forbes, 4.
41. Harrington, letter to Forbes.
42. Forbes, letter to Grove.
43. Forbes, Grove, Howell and Ankeny, *Report*, 33.
44. Ibid.
45. Ibid., 34.
46. Ibid.
47. Ibid.
48. Ibid.
49. Ibid.

50. "Identification of Bodies," *Ohio State Journal* (Columbus), B11.
51. Lawson, handwritten letter to the editor of the *Athens Messenger*.
52. "County Red Cross," *Sunday Messenger* (Athens, OH), 1.
53. "Aid Promptly Sent," *Glouster* (OH) *Press*, 1.
54. Walker, report to Forbes.
55. Morgan, letter to Forbes.
56. Tomlinson, memorandum to Forbes, 2.
57. Walker, report to Forbes, 1.
58. Tomlinson, memorandum to Forbes, 4.
59. Ankeny, "Story of the Shift," 1.
60. Walker, report to Forbes,2.
61. Ankeny, "Story of the Shift," 2.
62. Smith, "Report of Mine Explosion," 4.
63. Tomlinson, memorandum to Forbes, 4–5.
64. Walker, report to Forbes.
65. Forbes, Grove, Howell and Ankeny, *Report*, 34.
66. Ibid., 34–5.
67. "County Red Cross," *Sunday Messenger* (Athens, OH).
68. "Aid Promptly Sent," *Glouster* (OH) *Press*, 1.
69. "82 Dead Taken from Millfield Mine 6 Disaster," *Glouster* (OH) *Press*, 1.
70. Tomlinson, memorandum to Forbes, 5.
71. Ibid., 5.
72. Forbes, Grove, Howell and Ankeny, *Report*, 34.
73. Walker, report to Forbes, 2.
74. "Coal Co. Contributes," *Glouster* (OH) *Press*, 1.
75. Forbes, Grove, Howell and Ankeny, *Report*, 35.
76. Morgan, untitled report to Forbes.
77. Ray, "Notes," 9.

Chapter 4

78. "Funerals of Victims," *Athens (*OH) *Messenger*, 1.
79. "Cash Awards," *Sunday Messenger* (Athens, OH), 1.
80. "Mine Widows Given Awards," *Sunday Messenger* (Athens, OH), 11.

Chapter 5

81. Smith, "Report of Mine Explosion," 5.
82. Walker, report to Forbes, 1.
83. Morgan, untitled report to Forbes, 2.
84. "Tytus, Visiting Cleveland, Charters Plane," *Cleveland* (OH) *News*, 1.
85. Forbes, Grove, Howell and Ankeny, *Report*, 37–38.
86. Morgan, untitled report to Forbes, 4.
87. Walker, report to Forbes, 2.
88. Smith, "Report of Mine Explosion," 5–7.
89. Rice, memorandum to members of the Mine Safety Board, "Substitute for Portion of Page 40, Report on Sunday Creek No. 6 Mine," attachment A.
90. Smith, "Report of Mine Explosion," 9.
91. Confirmed in two sources: Owings, memorandum, 1; Tomlinson, memorandum to Forbes, 2.
92. Tomlinson, memorandum to Forbes, 2.
93. Confirmed in three sources: Owings, memorandum, 1; Ray, "Notes," 3; Forbes, Grove, Howell and Ankeny, *Report*, 14.
94. Forbes, Grove, Howell and Ankeny, *Report*, 14.
95. Owings, memorandum, 2.
96. Ibid.
97. Ibid.
98. Smith, "Report of Mine Explosion," 9–10.
99. Confirmed in two sources: Ray, "Notes," 8; Forbes, Grove, Howell and Ankeny, *Report*, 28.
100. Ray, "Notes," 11.
101. Tomlinson, memorandum to Forbes, 6–7.
102. Ray, "Notes," 1.
103. Ray, "Notes," 12.
104. Rice, memorandum to members of the Mine Safety Board, 4.
105. Owings, memorandum, 2.
106. *Report*, 20.
107. Forbes, Grove, Howell and Ankeny, *Report*, 5.
108. Ibid., 3.
109. Confirmed in two sources: Forbes, Grove, Howell and Ankeny, *Report*, 7; Tomlinson, memorandum to Forbes, 6.
110. Owings, memorandum, 1; Ray, "Notes," 3.
111. Tomlinson, memorandum to Forbes, 6.
112. Owings, memorandum, 2.

113. Tomlinson, memorandum to Forbes, 2.
114. Walker, report to Forbes, 2.
115. Ray, "Notes," 11.
116. Walker, report to Forbes, 2.
117. Tomlinson, memorandum to Forbes, 6–7.
118. Forbes, letter to Harrington, November 14, 1930.
119. Fene, letter to Hunter.

BIBLIOGRAPHY

Ankeny, M.J. "Story of the Shift That Went in at 7 O'Clock Thursday Morning to Find the Body of the Pumper." Department of Mines, Safety Division. Technical Information and Library, National Mine Health and Safety Academy. N.d.

Athens (OH) *Messenger*. "Disaster Sidelights." November 7, 1930, 9. https://newspaperarchive.com/athens-messenger-nov-07-1930-p-9/.

———. "Funerals of Victims Are Being Arranged All Over the District." November 7, 1930, 1. https://newspaperarchive.com/athens-messenger-nov-07-1930-p-1/

———. "Millfield Mine Disaster Traps Nearly 150 Miners: Lives of Miners Are in Grave Peril Due to Falling of Shale." November 5, 1930, 1.

———. "Millfield Mine of Poston Co. Is Taken Over." August 1, 1929, 1. https://newspaperarchive.com/athens-messenger-aug-01-1929-p-1/.

———. "Pioneer Coal Man Passes." August 23, 1923, 1. https://newspaperarchive.com/athens-messenger-aug-23-1923-p-1/.

———. "Poston Mine Sets Record for Day: 1,701 Tons Produced at Millfield Workings on Thursday, Report Shows." February 9, 1930, 12. https://newspaperarchive.com/athens-sunday-messenger-feb-09-1930-p-12/.

———. "Store Is Sold: Sunday Creek Coal Co. Buys Columbian Store at Millfield." January 31, 1930, 5. https://newspaperarchive.com/athens-messenger-jan-31-1930-p-5/.

———. "Three Shifts of Federal State and County Men to Probe Depths of the Workings." November 7, 1930, 1.

Attorney General. "Syllabus." 1930, 1,083–91. chrome-extension://efaidnbmnnnibpcajpcglclefindmkaj/https://www.ohioattorneygeneral.gov/getattachment/1608fffe-7be9-43d3-9fae-432b893b513f/1931-3510.aspx.

Bancroft, Thomas B. *Eleventh Annual Report of the Chief Inspector of Mines to the Governor of the State of Ohio for the Year 1885*. Columbus, OH: Westbote, State Printers, 1886.

———. *Twelfth Annual Report of the Chief Inspector of Mines to the Governor of the State of Ohio for the Year 1887*. Columbus, OH: Westbote, State Printers, 1888.

Biddison, Elmer G. *Twenty-Eighth Annual Report of the Chief Inspector of Mines to the Governor of the State of Ohio for the Year 1902*. Springfield, OH: Springfield Publishing, State Printers, 1903.

———. *Twenty-Fifth Annual Report of the Chief Inspector of Mines to the Governor of the State of Ohio for the Year 1899*. Columbus, OH: Fred J. Heer, n.d.

———. *Twenty-Fourth Annual Report of the Chief Inspector of Mines to the Governor of the State of Ohio for the Year 1898*. Columbus, OH: Westbote, 1899.

———. *Twenty-Ninth Annual Report of the Chief Inspector of Mines to the Governor of the State of Ohio for the Year 1903*. Springfield, OH: Springfield Publishing, State Printers, 1904.

———. *Twenty-Seventh Annual Report of the Chief Inspector of Mines to the Governor of the State of Ohio for the Year 1901*. Columbus, OH: Fred J. Heer, State Printer, 1902.

———. *Twenty-Third Annual Report of the Chief Inspector of Mines to the Governor of the State of Ohio for the Year 1897*. Norwalk, OH: Laning Printing, 1898.

Brooks, Tonya. "Millfield Man Says He Was 'Lucky' During Fabled 1930 Mine Explosion." *Athens News*, December 8, 2008, 18.

Business Details and Filings: State of Ohio Government. "Millfield Coal and Mining." https://businesssearch.ohiosos.gov?=businessDetails/28725.

———. "Sunday Creek Coal Company." https://businesssearch.ohiosos.gov?=businessDetails/85685.

Cleveland (OH) *News*. "Mrs. W.E. Tytus, Visiting Cleveland, Charters Plane." November 6, 1930, 1.

Columbus (OH) *Citizen*. "Mine Explosions Are Held to Be Preventable." November 7, 1930. (A retyped copy of the article was provided to the author as part of materials requested from Melody Bragg, a technical information specialist for the Technical Information Center and Library, National Mine Health and Safety Academy, on March 25, 2011. The retyped version does not include a volume or number. The original newspaper could not be located in a search of *Chronicling America* or

Newspaper Archives and did not appear in the catalog of the Ohio History Center, the online holdings of the *Columbus Citizen Journal/Newspaper*, or at the Columbus Public Library.)

———. "U.S. Official Sees Negligence." November 6, 1930.

Cross, Roy. "New Mine Tragedy Recalls Millfield's." *Athens* (OH) *Messenger*, November 24, 1968, E1.

———. "Tragedy at Millfield." *Athens* (OH) *Messenger*, November 2, 1997, D1.

Crowell, Douglas L. *Bulletin 72: History of the Coal-Mining Industry in Ohio.* Columbus, Ohio: ODNR, 1995.

———. "Death Underground: The Millfield Mining Tragedy." *Timeline* 14, no. 5 (September/October 1997): 42.

———. "Millfield Tragedy Revisited." *Ohio Geology* (Fall 1995): 5.

Davis, Howard J. "Testimony of Witnesses Inquest Over the Dead Body of Victims of Mine Number 6, Millfield Disaster." Coroner's inquest. November 12, 1930. 4 pages.

Democrat-Sentinel (Logan, OH). "81 Miners Are Found Dead." November 6, 1930, 1.

Earich, Adolph. "Miners." Handwritten text. 1975.

Ericksen, Annette G., et al. "Sunday Creek Coal Company Mine No. 6 (Millfield Mine). RR1 CR27 (Millfield Road) East Millfield, Athens County, Ohio." Historic American Engineering Record, OH-139. 2013. https://www.loc.gov/item/oh2011/.

Fay, Albert H. "Coal-Mine Fatalities in the United States 1870–1914." Bulletin 115. 1916. https://digital.library.unt.edu/ark:/67531/metadc12321/m2/1/high_res_d/Bulletin0115.pdf.

Feehan, Francis. Letter to C.B. Mead. November 10, 1930.

Fene, W.J. Letter to Van B. Hunter. November 12, 1930.

———. Telegram to D. Harrington. November 5, 1930, 2:54 p.m.

———. Telegram to D. Harrington. November 6, 1930, 11:00 a.m.

Forbes, J.J. Letter to D. Harrington. December 1, 1930.

———. Letter to D. Harrington. December 9, 1930.

———. Letter to D. Harrington. November 14, 1930.

———. Letter to D. Harrington. September 12, 1931.

———. Letter to E.W. Smith. December 1, 1930.

———. Letter to G.W. Grove. December 9, 1930.

———. Letter to H.T. Bannister. December 4, 1930.

———. Letter to L.E. Geohegan, Gulf States Steel Company. November 17, 1930.

———. Telegram to D. Harrington. November 10, 1930, 1:20 pm.
———. Telegram to U.S. Bureau of Mines. November 6, 1930, 8:41 p.m.
———. Telegram to U.S. Bureau of Mines. November 7, 1930, 6:21 p.m.
———. Telegram to W.J. Fene. November 9, 1930, 12:35 a.m.
Forbes, J.J., G.W. Grove, S.P. Howell and M.J. Ankeny. *Report of Gas and Dust Explosion in Mine No. 6, Sunday Creek Coal Company, Millfield, Ohio, November 5, 1930.* Department of Mines, Safety Division. Technical Information and Library, National Mine Health and Safety Academy.
Glouster (OH) *Press*. "Aid Promptly Sent into Stricken District." November 13, 1930, 1.
———. "Coal Co. Contributes $9000 to Burial Fund." November 13, 1930, 1.
———. "82 Dead Taken from Millfield Mine 6 Disaster: Worst in History Ohio Mines." November 13, 1930, 1.
Green and White (Ohio University). "Praiseworthy." November 7, 1930, 2. https://media.library.ohio.edu/digital/collection/studentnewspapers/id/19480/rec/7.
Harrington, D. Letter to J.J. Forbes. November 28, 1930.
———. Telegram to J.J. Forbes. November 7, 1930, 12:34 p.m.
Harris, C.H. "Scene of Millfield Mine Disaster." *Athens* (OH) *Messenger*, July 8, 1952, 7.
Harrison, George. *Thirty-Eighth Annual Report of the Chief Inspector of Mines to the Governor of the State of Ohio for the Year Ending December 31, 1908.* Springfield, OH: Springfield Publishing, State Printer, 1909. https://books.google.com/books.
———. *Thirty-Fifth Annual Report of the Chief Inspector of Mines to the Governor of the State of Ohio for the Year Ending December 31, 1909.* Springfield, OH: Springfield Publishing, State Printer, 1910. https://books.google.com/books.
———. *Thirty-First Annual Report of the Chief Inspector of Mines to the Governor of the State of Ohio for the Year Ending December 31, 1905.* Columbus, OH: Fred J. Heer, State Printer, 1906. https://books.google.com/books.
———. *Thirty-Fourth Annual Report of the Chief Inspector of Mines to the Governor of the State of Ohio for the Year Ending December 31, 1912.* Springfield, OH: Springfield Publishing, State Printer, 1913.
———. *Thirty-Second Annual Report of the Chief Inspector of Mines to the Governor of the State of Ohio for the Year Ending December 31, 1906.* Columbus, OH: F.J. Heer, State Printer, 1907. https://books.google.com/books.
———. *Thirty-Sixth Annual Report of the Chief Inspector of Mines to the Governor of the State of Ohio for the Year Ending December 31, 1910.* Columbus, OH: F.J. Heer, State Printer, 1911. https://books.google.com/books.

———. *Thirty-Third Annual Report of the Chief Inspector of Mines to the Governor of the State of Ohio for the Year Ending December 31, 1907.* Columbus, OH: F.J. Heer, State Printer, 1908. https://books.google.com/books.
Haseltine, Robert M. *Eighteenth Annual Report of the Chief Inspector of Mines to the Governor of the State of Ohio for the Year 1892.* Columbus, OH: Westbote, State Printer, 1893.
———. *Seventeenth Annual Report of the Chief Inspector of Mines to the Governor of the State of Ohio for the Year Ending November 15, 1891.* No publisher information.
———. *Twenty-First Annual Report of the Chief Inspector of Mines to the Governor of the State of Ohio for the Year 1895.* Columbus, OH: Westbote, 1896. https://books.google.com/books.
———. *Twenty-Fourth Annual Report of the Chief Inspector of Mines to the Governor of the State of Ohio for the Year 1898.* Columbus, OH: Westbote, 1899. https://books.google.com/books.
———. *Twenty-Second Annual Report of the Chief Inspector of Mines to the Governor of the State of Ohio for the Year 1896.* Norwalk, OH: Laning Printing, 1896. https://books.google.com/books.
———. *Twenty-Third Annual Report of the Chief Inspector of Mines to the Governor of the State of Ohio for the Year 1897.* Norwalk, OH: Laning Printing, 1896. https://books.google.com/books.
Humphrey, H.B. "Historical Summary of Coal-Mine Explosions in the United States." *U.S. Bureau of Mines Information Circular* 7900 (1959): 5–6, 38–41, 124, 126–27, 168–74, 236–40. Reprint from the collection of the University of Michigan Library.
Knox, F.S., Sunday Creek Coal Company. Letter to George K. Smith. February 24, 1931.
Lawson, William. Handwritten letter to the editor of the *Athens Messenger.* September 28, 1988.
Loughridge, Susan, ed. *Millfield Mine Disaster: November 5, 1930.* Millfield, OH: Millfield Mine Memorial Committee, 1977.
Mackie, John. Handwritten letter to Roy Cross. November 3, 1997.
McDonald, H.L. "The Millfield Mine Explosion." Self-published brochure, November 5, 1930.
Millfield Christian Church. "1994 Commemoration of Millfield Mine Disaster." Millfield, OH. November 5, 1994.
"Millfield Mine Disaster Correction." *Ohio Geology* (Winter 1996): 7.

Morgan County (McConnelsville, OH) *Democrat*. "Modest Scot Saves Eighty Men from Mine Disaster." November 13, 1930, 7. https://chroniclingamerica.loc.gov/lccn/sn87075008/1930-11-13/ed-1/seq.

Morgan, Raymond A. Letter to J.J. Forbes. November 25, 1930

———. Untitled report to J.J. Forbes. December 26, 1930. 5 pages.

National Register of Historic Places Inventory. "Sunday Creek Coal Company Mine No. 6." Nomination Form. January 31, 1978; entered May 23, 1978. https://s3.amazonaws.com/NARAprodstorage/lz/electronic-records/rg-079/NPS_OH/78002005.pdf.

"1989 Commemoration of Millfield Mine Disaster." Millfield, OH. November 4, 1989.

Ohio State Journal (Columbus). "Identification of Bodies Began Late Wednesday." November 7, 1930, B11.

Owings, C.W. Memorandum to Members of the Safety Division, U.S. Department of Commerce. February 19, 1931.

Peters, William E. *Athens County, Ohio*. Vol. 1. N.p.: self-published, 1947.

———. "Range 14, Town 11 (Trimble Township, Athens County, Ohio: 65." W.E. Peters Papers, Range Books, Volume 21. https://media.library.ohio.edu/digital/collection/wepeters/id/4644/rec/25.

Ray, Frank A. (consulting mining engineer). "Notes Taken from U.S. Bureau of Mines Report on Their Investigation, Rescue Work, etc., on the Explosion in the Sunday Creek Coal Company's Mine, Poston No. 6 at Millfield, Athens County, Ohio Which Occurred November 5th, 1930." 14 pages.

Rice, George S. Letter to J.T. Ryan. January 5, 1931.

———. Memorandum to members of the Mine Safety Board. January 30, 1931.

Roy, Andrew. *A History of the Coal Miners of the United States from the Development of the Mines to the Close of the Anthracite Strike of 1902, Including a Brief Sketch of Early British Miners*. Columbus, OH: J.L. Trauger Printing Company, n.d. (This book would have been published sometime between 1902 and 1905, since a revised edition was completed in 1906.)

Schatenstein, Anne. "76 Are Known Dead: County Coal Mine." *Columbus* (OH) *Evening Dispatch*, November 6, 1930, 1.

Sharp, Harry. "Climax to Come First of April in Mine Strike." *Canton* (OH) *Daily News*, February 13, 1928, 1. https://newspaperarchive.com/canton-daily-news-feb-13-1928-p-1/.

Smith, E.W. "Report of Mine Explosion Which Occurred at 11:45 A.M., November 5, 1930, in Mine No. 6 of the Sunday Creek Coal Company at

Millfield, Ohio, in Which 73 Employes, 5 Officials of the Company and 4 Visitors Were Killed." N.d. (This document was provided to the author as part of materials requested from Melody Bragg, a technical information specialist for the Technical Information Center and Library, National Mine Health and Safety Academy, on March 25, 2011. The item states, "This report was prepared by E.W. Smith, Chief of Division of Mines, Cambridge, Ohio, but is not signed by Smith or any other person.")

———. Telegram to J.J. Forbes. November 13, 1930, 4:41 p.m.

Smith, George K. Letter, with attached letter and memorandum from F.S. Knox to F.A. Ray. March 3, 1931.

Smith, S. Winifred. "The Millfield Mine Disaster." *Museum Echoes*, August 1959, 59.

Sunday Messenger (Athens, OH). "Cash Awards Made to Mine Victim Families: Payments by Week Are to Be Made to Widows of Miners." November 16, 1930, 1. https://newspaperarchive.com/athens-sunday-messenger-nov-16-1930-p-11/.

———. "Committee Elects Ayers President." February 2, 1975, A8.

———. "County Red Cross Carrying Out Detailed Relief Work." November 9, 1930, 1. https://newspaperarchive.com/athens-sunday-messenger-nov-9-1930-p-1/.

———. "Mine Widows Given Awards." November 23, 1930, 11. https://newspaperarchive.com/athens-sunday-messenger-nov-23-1930-p-27/.

———. "Stork Arrives in a Sorrowing Home." November 9, 1930, 8. https://newspaperarchive.com/athens-sunday-messenger-nov-9-1930-p-1/.

———. "The Sunday Creek Coal Company Underwrote the Funeral Expenses Friday of the 82 Victims of the Mine Disaster." November 9, 1930, 1. (Single page found in collection of the Southeast Ohio History Center on September 1, 2011.)

Tomlinson, Walter H. Memorandum to J.J. Forbes. November 20, 1930.

Tourjee, William. "Millfield Mine Disaster Remembered." *Athens* (OH) *Messenger*, November 4, 1974, 2.

United States Department of Labor, Mine Safety and Health Administration. "Coal Fatalities for 1900 through 2022." https://arlweb.msha.gov/stats/centurystats/coalstats.asp.

Walker, Dan. Report to J.J. Forbes. February 28, 1931.

Watkins, Damon D. *Keeping the Home Fires Burning: A Book About the Coal Miner*. 2nd ed. Columbus: Ohio Company, 1937.

Zern, E.N. *Coal Miners' Pocketbook: Principles, Rules, Formulas and Tables*. New York: McGraw-Hill, 1928.

INDEX

C

D

E

F

G

H

I

J

N

O

P

Q

R

S

T

U

V

W

Y

ABOUT THE AUTHOR

Ron Luce earned a BA from State University of New York, College at Brockport, and an MA and PhD from Ohio University. He taught for many years at Hocking College, Ohio University and Virginia Tech. In his professional life, he wrote numerous professional articles, edited a professional journal, served on boards of prominent organizations and made many presentations at the local, state and national levels. Upon his retirement from teaching, he served as the executive director for the Athens County Historical Society and Museum, where he expanded his knowledge and appreciation of local history and worked to educate the public about the county's rich history through presentations, publications and supporting other local historians, at times serving as an editor for their books. He has written plays, short stories, poems and novels. His newest novel, *Star Late Rising*, will be released in April 2024.

Over his many years, Luce has received numerous awards and recognition for his teaching, writing, artistic work and civic engagement.